P9-CCD-070

GREEN INTENTIONS

Creating a Green Value Stream to Compete and Win

GREEN INTENTIONS

Creating a Green Value Stream to Compete and Win

BRETT WILLS

CRC Press is an imprint of the
Taylor & Francis Group, an **informa** business
A PRODUCTIVITY PRESS BOOK

Productivity Press
Taylor & Francis Group
270 Madison Avenue
New York, NY 10016

© 2009 by Taylor and Francis Group, LLC
Productivity Press is an imprint of Taylor & Francis Group, an Informa business

No claim to original U.S. Government works

Printed in the United States of America on acid-free paper
10 9 8 7 6 5 4 3 2 1

International Standard Book Number: 978-1-4200-8961-5 (Paperback)

This book contains information obtained from authentic and highly regarded sources. Reasonable efforts have been made to publish reliable data and information, but the author and publisher cannot assume responsibility for the validity of all materials or the consequences of their use. The authors and publishers have attempted to trace the copyright holders of all material reproduced in this publication and apologize to copyright holders if permission to publish in this form has not been obtained. If any copyright material has not been acknowledged please write and let us know so we may rectify in any future reprint.

Except as permitted under U.S. Copyright Law, no part of this book may be reprinted, reproduced, transmitted, or utilized in any form by any electronic, mechanical, or other means, now known or hereafter invented, including photocopying, microfilming, and recording, or in any information storage or retrieval system, without written permission from the publishers.

For permission to photocopy or use material electronically from this work, please access www.copyright.com (http://www.copyright.com/) or contact the Copyright Clearance Center, Inc. (CCC), 222 Rosewood Drive, Danvers, MA 01923, 978-750-8400. CCC is a not-for-profit organization that provides licenses and registration for a variety of users. For organizations that have been granted a photocopy license by the CCC, a separate system of payment has been arranged.

Trademark Notice: Product or corporate names may be trademarks or registered trademarks, and are used only for identification and explanation without intent to infringe.

Library of Congress Cataloging-in-Publication Data

Wills, Brett.
Green intentions : creating a green value stream to compete and win / Brett Wills.
p. cm.
Includes bibliographical references and index.
ISBN 978-1-4200-8961-5
1. Green marketing. 2. Environmental economics. I. Title.

HF5413.W55 2009
658.4'083--dc22 2009001497

Visit the Taylor & Francis Web site at
http://www.taylorandfrancis.com

and the Productivity Press Web site at
http://www.productivitypress.com

To My Mother for her lifelong love, support and encouragement.

A Special Thanks:

To Philip Ling for making it "click" and proving that
one person's passion can make a difference.

To Bob Kerr for all the mentorship, guidance and
providing the opportunity to make a difference.

Contents

Foreword

The Greening of Manufacturing

If there is a major surprise at this time of heightened awareness about the importance of energy efficiency, the potential impacts of climate change, and the effects of environmental degradation, it is that relatively little in the way of thought leadership has been devoted to what we now commonly call green manufacturing. Even less thought and hard analysis has been devoted to developing a systematic operational approach that is useful to manufacturers in managing their relationship with the environment and that also makes sense from a business perspective.

Yet, we know that customers are increasingly searching for products that leave a minimal environmental footprint. Customers, governments, standards organizations, investors, and many other stakeholders are requiring manufacturers and their suppliers to meet new and much more stringent product, process, and environmental management standards. And, their expectations of the environmental responsibilities and liabilities of manufacturers now extend to full life-cycle management of products from inception, through development and production, to customer use and final disposal.

On the other hand, green manufacturing brings with it new business opportunities. Manufacturers should be looking for systematic ways of eliminating energy and other environmental wastes, cutting unnecessary materials and regulatory overheads, and generating cash flow—all of which are even more important at a time of economic downturn and weak profit margins. There are new market opportunities for manufacturers as well—in supplying new products, solving environmental and technological problems, and introducing product management services to customers around the world. Green implies a level of product and service quality, specialization, customization, and sophistication that differentiates manufacturers who adopt its precepts from the commodity production and cut-throat cost competition common in the manufacture and sale of standard products.

The greening of manufacturing is no fad. It's a global phenomenon. It's part of a major shift in the products, technologies, production processes, services, and markets that is changing the very nature of manufacturing around the world.

We frequently hear more about the need for greater energy efficiency and the need to reduce greenhouse gas emissions to meet new regulatory obligations and address the issue of climate change. Canadian manufacturers have a good story to tell in this regard. Between 1990 and 2007, manufacturers across Canada reduced their greenhouse gas emissions by about 10 percent. They actually surpassed Canada's Kyoto target of a 6 percent emission reduction. They succeeded in doing so because of improved energy efficiency efforts (accounting for about half of emission reductions); the introduction of new, improved, and less carbon-intensive industrial processes (cutting emissions by a further 30 percent); and switching from fossil fuels to less carbon-intensive forms of energy (another 20 percent).

Now manufacturers are being expected to go even further and increase their rate of emission reduction over the next 15 years. Can they do this? Well, that depends on whether they have the environmental monitoring, management, and financing systems in place to achieve continuous improvement in energy management and emission reductions. It also depends on their investment in and adoption of new, more energy efficient and less-carbon intensive production technologies.

At least four lessons can be learned from the emission-reduction record of Canadian manufacturers since 1990. The first is that energy efficiency improvements make good sense for any business looking to cut energy costs. Energy costs are not fixed costs, and they can be managed.

The second lesson is that lean improvements do contribute to energy efficiency, carbon emission reductions, and reducing environmental waste. By eliminating non-value-adding activities, businesses reduce their energy demand. There is a direct connection between "lean" and "green.".

The third lesson, though, is that a more systematic approach to energy efficiency, the reduction of carbon, and the elimination of environmental waste is needed in order to sustain significant operating improvements over a period of time. In my mind, this is an important lesson because it is what really connects lean and green. Remember, lean is a business philosophy. It is about identifying what customers value and eliminating those non-value-adding activities that are the waste in any process or business system. But, lean also tells us to question why—repeatedly—to get to the heart of understanding processes. It is not only about identifying where the seven wastes are in a process; it is about identifying all sources of waste. Clearly energy and environmental waste are very significant by-products of non-value-adding activities for manufacturers today. It would make sense, therefore, that the approach to identifying and eliminating energy and environmental waste should be similar to that of eliminating other wasteful activities. Lean principles and methodologies work to "green-out" business processes as well. Let's just start by including environmental waste as part of the *muda* that needs to be eliminated.

The fourth lesson that can be drawn from the emission reduction achievements of Canadian manufacturers over the past two decades is the importance of investment in new technologies. Actually, what really counts is taking old machinery and equipment out of commission and replacing them with new, more productive technologies that are also energy efficient and more environmentally friendly at the same time. What is important here is that both productivity and environmental benefits must be translated into financial returns on investment. As with other lean improvements, achieving positive financial returns at the end of the day is a precondition for sustaining process improvements, or any innovation.

Each of these lessons comes down to one key point: Energy and environmental management can and should be a part of business planning for efficiency improvement, product innovation, and market development. It must make sense from a financial perspective. And, as with all business plans, the challenge comes down to execution, to making objectives and good intentions operational.

This is why I welcome a guide that employs lean principles and methodologies to specifically target environmental waste. Good environmental management is more than a by-product of lean, but there is much in the lean lexicon that can be applied to eliminating environmental waste. And integrating environmental management into proactive product and process innovations can help to grow business opportunities in the future.

Brett Wills is a thoughtful leader who shows how lean can be applied to green. Read on.

Jayson Myers
President
Canadian Manufacturers & Exporters

Introduction

The Greening of Commerce

These days, you can't turn on a television or radio, search the Internet, or pick up a newspaper without hearing about the environment. News of a natural disaster, the extinction of a species, rising temperatures, melting glaciers—there is no escaping it no matter where you turn. Although this new focus on the environment is a good thing, one that brings to light the seriousness of the fact that the planet may be in dire straits, there is a downside. Unfortunately, we are inundated with so many issues, concerns, and crises that we may have a hard time weeding out the real issues and understanding how to deal with them, so that real progress can be made.

Whatever Happened to the Ozone Layer?

Only a few years ago, the media frequently mentioned a "hole" in the ozone layer, yet today, we never hear about it. Why? Well, because we did something about it. Because enough environmental activists and politicians understood the issue and focused on a way to deal with it, they developed the Montreal Protocol of 1987, which worked to eliminate CFCs (chlorofluorocarbons—gaseous materials that rise up to the ozone level and wear it away). CFCs are no longer being produced and will soon not be in use at all. This change has slowed the growth of the hole in the ozone layer, and over time, the ozone layer will regenerate.

This example serves as inspiration: We *can* deal with and develop solutions for the environmental issues we face today. I believe that we have the ability not only to develop the solutions, but also to implement those solutions fast enough to make a difference. However, forced compliance to environmental regulations like that of the Montreal Protocol can achieve only so much. If we rely only on these legislative mechanisms, we will not make it. The answer to truly addressing the pressing environmental issues we face today lies in the greening of enterprise. The aggregate environmental impact of global enterprises is causing a majority of the environmental issues facing us today. By greening enterprise, we can eliminate a majority of these environmental issues. Unfortunately, starting in the 1970s, green developed a reputation of being costly and of requiring specialized knowledge and expertise, with no immediate benefits or results beyond the obvious environmental ones. This mindset is beginning to change. People are starting to see that by going green, not only will they save the environment, but they will also reap tremendous benefits on the bottom line. Businesses are also seeing the environmental writing on the wall: businesses that don't go green will be left behind by competitors who do, or will be forced by governing agencies to go green. Forget for a moment that the Earth will eventually force a green way of life by not providing additional natural resources, by changing climates, and so on. Forget that, because although going green is the right thing to do, it's also the profitable, competitive approach to take.

The Business Case for Environmental Sustainability

The economic benefits of going green are tremendous, and every business that has yet to move in this direction has an untapped gold mine right under its feet. Those that have started to move in this direction know the size of the gold mine and are starting to figure out ways to mine it. The economic or business benefits to going green are numerous but, for simplicity, I have grouped them into six major categories:

- Cost savings
- Increased customer loyalty and attraction
- Increased employee attraction and retention
- Ability to grow
- Innovation and development of new technologies
- Increased profit and shareholder value

Cost Savings

Perhaps the most talked about business benefit of going green is the potential for cost savings on everything from reduced raw material costs to reduced operational and administrative costs. This is true, and the logic is simple: if you use fewer raw materials or reuse your raw materials—both of which provide great environmental benefits—you will also save money. If you use less energy and less water, create less garbage, and reduce transportation distances, you will save on operational and administrative costs. Some companies already are harvesting these savings. According to a 2008 study by *The Economist* magazine entitled "Doing Good: Business and the Sustainability Challenge," DuPont has cut costs by $2 billion since 1990 through energy reduction initiatives alone; 3M has saved $82 million between 2001 and 2005 and reaped another $10 million in savings in 2006. Although these are exceptional circumstances, there are things you can do quickly and easily with little cost or effort that produce substantial savings. Coming up with solutions oftentimes is rather easy, since many solutions are already out there and can be mimicked or modified to fit individual companies. The challenge is to be able to spot the opportunities for savings (this requires a shift in thinking) so that you can find the best-practice solution, and then having the discipline to see the project through.

Increased Customer Loyalty and Attraction

With global competition now being the norm, we know how hard it can be not only to attract new customers and break into new markets, but also to keep the customers and market share we already have. By adopting environmentally sustainable practices, you can deal with this challenge successfully. Although it is harder to measure how many customers you have kept or attracted because of green initiatives, you can be sure that your efforts will be well worth it. A February 2008 survey conducted by Cone, Inc., in conjunction with the Boston College Center for Corporate Citizenship confirms this claim. According to the 1,080 adults surveyed, 59 percent were concerned about the environmental impact of their consumption and were changing buying habits in order to lessen the impact. Sixty-six percent of the respondents indicated that they consider the environmental impact of their purchases. Sixty-eight percent of those surveyed also said that if the company had a strong reputation for environmental commitment, it positively influenced their decision to buy that company's product or service.

In addition to this survey, Cone, Inc., conducted another survey of "millennials," those born between 1979 and 2001, to find that this group had even stronger results to support that claim. This survey found that 83 percent of millennials will trust a company more if it is environmentally responsible. Sixty-nine percent consider a company's social and environmental commitment when deciding where to shop. Eighty-nine percent will switch from one brand to another if the second is associated with a good cause, and 74 percent will pay more attention to a company's message if the company has a deep commitment to a cause. What does all this mean? If you are going green and continue to be committed to going green you will have a much better chance of keeping existing customers and attracting new customers away from competitors that are lagging. So the sooner you start, the faster you move, and the more committed you are, the better.

Increased Employee Retention and Attraction

We all know the importance of keeping and attracting good employees, from a development and growth perspective as well as from a cost perspective, as in the costs associated with rehiring and retraining. It is also no surprise that global competition not only jeopardizes our customer base, but also impacts our ability to keep and attract good employees. Once again, green or environmental solutions will help address this challenge. Of the 28 percent of respondents from Cone, Inc.'s, millennial survey who worked full time, 79 percent wanted to work for a company that cares about how it impacts or contributes to society. Sixty-four percent said that their company's social and environmental activities make them feel loyal to their company, and an astonishing 68 percent said they would refuse to work for a company that is not socially and environmentally responsible. It is no secret that there is a shortage of skilled workers, especially in the manufacturing industry within North America, and considering that millennials are going to be taking over from the aging baby boomers, this is important information. The results are cut and dry: To keep your good employees and attract the best and brightest, you'd better be moving on with going green … and fast, before your competitors get ahead of you.

Ability to Grow

By "ability to grow," I mean the ability to physically support the increase in the size of your business given the shrinking supply of and increased demand for natural resources. For example, if your products are petroleum-based, like plastics for instance, your cost-cutting efforts will only be eaten away by the increase in petroleum prices. On the other hand, if you are focused on the environmental impacts of your products, you will find alternative, environmentally friendly substitutes for plastics, eliminating the inevitable material cost increases for the remainder of the product's life. Or, if you are using methods to recover used materials so that you can use them again, even better. Another example is operations that are energy intensive. If you do not focus on energy conservation and alternative clean energy sources, you are assured of never-ending increases in energy costs. On the other hand, by focusing on reducing your environmental impact, you can find ways to lower energy costs and move toward self-harvested clean energy that will be free after you pay off the initial investment. If you are overharvesting materials or natural resources faster than they are regenerating or using materials that are not regenerating at all, there will eventually be no more materials, and not only will you be unable to grow, but you won't be able to produce what you are making now. Also, by going green you increase your ability to get financing. Many banks are now using environmental impact as a criteria for lending and we all know the importance of securing financing when looking to grow. In addition to this there are stricter laws and legislation with regards to environmental impact coming down the pipe and failure to comply will certianly inhibit and limit growth. Greening your operations will help you ensure your ability to grow. For all these reasons, it is imperative that a business interested in sustaining its growth start the journey toward environmental sustainability before it's too late.

Innovation and Development of New Technologies

With a focus on going green, you will constantly be looking at how to do things differently and better in order to reduce your impact on the environment. By doing things differently and better, you are being innovative—thus, going green drives innovation.

Being innovative with a green motivation—as opposed to any other motivation—not only gives you direct environmental and cost savings, but also increases productivity, reduces lead times, increases capacity, and so on, leading to more savings and benefits, and resulting in more value for your customers. For example, if you are using less material, it takes less time to process that material, resulting in reduced process costs, reduced lead times, and increased capacity. If you use less transportation, you will reduce your lead times. If you generate less garbage, and therefore have less to move around and deal with, you increase your capacity to build product and contribute to shorter lead times.

Also, by focusing on green, you will start to look outside your four walls at how you can help your customers reduce their environmental impact through the use of your products and services. The market for new green technologies, both on business-to-business and business-to-consumer levels, is growing exponentially and will only continue to grow as businesses and consumers worldwide look for ways to lessen their environmental impact. As you drive the development of new and greener technologies, you provide great value to your customers by helping them reduce their environmental footprint, and you will also ensure that you are able to meet the demands of your customers now and in the future.

Increased Profit and Shareholder Value

By going green, you can realize savings and increased profits beyond your wildest imagination. In fact, companies can increase their profits by 35 percent or more, often in under five years. Research done by Bob Willard (an internationally renowned leader in sustainability strategies, author of numerous books on the business case for sustainability, and veteran of implementing real-life solutions), shows that a typical large enterprise can yield, on average, a 38 percent increase in profit within five years. His research also shows that smaller enterprises can yield even higher numbers, often over 50 percent. This increase in profit typically comes from attracting better and more productive employees; reducing materials, manufacturing, and operating costs; and increasing the revenues from new and existing customers. But proof of this comes not from research or studies, but from the cold, hard financial results. The companies listed in the Dow Jones Sustainability Index, an index that tracks the financial performance of the leading sustainability-driven companies worldwide, constantly outperform the rest of the market. Such performance will capture the attention of investors, as shareholders tend to invest in the companies that provide the greatest value and return on investment.

The challenge for many companies has been the lack of a complete, easy-to-use road map for greening an organization. There are a number of ideas, concepts, and tips for going green—so many that it is confusing. This book, and the green value stream approach detailed within it, provides a clear understanding of what it means to go green, as well as the road map for getting there. By understanding and following the Lean & Green process within your organization, you will be able to quickly reap the benefits of going green. Taking this step will catapult your organization ahead of your competitors, allow you to experience tremendous growth, and solidify the foundation of your organization.

What You'll Find in This Book

For the most part, people understand many of the challenges facing the greening of commerce. It isn't rocket science, but rhetoric can cloud the issues as well as the path that leads to environmental sustainability. For this reason, Appendix A provides you a "green dictionary," to which you can refer whenever a term requires a deeper definition than the text in the rest of this book offers. Appendix B offers another chance to come up to speed, by providing a quick primer on the environmental issues of today. Flip to Appendix B if you are searching for additional information on the most important environmental issues, their associated impacts, and the end goal in dealing with those issues. (The end goal is environmental sustainability, but that notion often gets lost.) Appendix C lists additional Web-based resources that will help you understand the current state of the environment and find the solutions to the issues and challenges that arise on your way to environmental sustainability.

Green Intentions takes you step-by-step in easy to understand language, through a process that greens your value streams by eliminating the seven major environmental wastes. This is your road map to environmental sustainability.

Here is a look at how this book is organized, chapter by chapter:

- **Section I, Going Green:** Chapter 1 shows how the green value stream approach, also known as Lean & Green, is based on the lean manufacturing principles, provides a dynamic, proven, and successful approach to going green. It also defines each of the seven green wastes you will work to minimize and eventually eliminate. Chapter 2 explains the overall green value stream process, provides guidance on implementing it in your organization, and shows how to map your green value streams.

- **Section II, The Seven Green Wastes:** Chapters 3 through 9 provide a detailed explanation of each of the seven green wastes and the "how-to," step-by-step process for minimizing and eliminating each of these wastes. Each chapter includes real-life examples illustrating the tremendous environmental and economic benefits associated with moving toward the elimination of each waste.
- **Section III, Conclusion and Appendices:** Chapter 10 offers a conclusion that briefly sums up this book. Appendix A provides you with a "green dictionary" that contains definitions of the most current and common terms associated with the green movement. Appendix B provides a number of Web links and other resources that will help you along your journey toward environmental sustainability. Appendix C is an environmental primer that clears through the rhetoric and confusion surrounding the state of the environment today and the end goal of sustainability, providing a clear picture of what is going on with the environment and what the end goal of environmental sustainability really is. Finally, Appendix D contains blank copies of the seven waste elimination worksheets used throughout the book.

GOING GREEN

I

Chapter 1

From Lean to Green: Green Value Stream Thinking

With all the different industries and businesses in the world doing so many different things, it would be nearly impossible to develop an exhaustive list of what each has to do in order to be greener. Instead, drawing on Toyota's principles of lean production, the trick is to follow a systematic process to achieve an end result, one that will vary from company to company and industry to industry. The systematic process can be applied to almost any business or organization, making it dynamic, but each can go green in the best way for that company.

With lean, the trick is to shift your thinking so that you're seeing every portion of your process from the perspective of the customer—what non-value-adding activities (waste) can be eliminated, based on whether they add anything to the customer's experience or satisfaction. This same way of thinking can be applied to green, with a slightly different perspective. Instead of seeing things from the perspective of the customer, you learn to see from the perspective of *the environment*. By shifting your thinking in this way, applying existing green technologies, techniques, and approaches, and using a proven framework or philosophy, you have a clear road map or process to go green.

In essence, you're looking at all the activities in the value stream or operation of a business from the perspective of the environment, a green value stream (GVS), if you will. As with lean, you use a set of criteria to check against the activities of your operation also known as your value stream. With lean, the criteria are based on what your customer does not perceive as positive, good, or valuable and, therefore, would not want to pay for—in other words, all the wasteful operations, processes, or things that the customer would not want. These seven lean wastes are famously known as:

- Inventory
- Movement
- Defects
- Transportation
- Overproduction
- Excess processing
- Waiting

By shifting your thinking to look at things from the customer's perspective, identifying each of these wastes within your value stream, and following the waste elimination process, you can effectively eliminate them. This lean process with its tools and techniques also optimizes value-adding activities, so that the customer is getting the most out of those activities. The result is reduced costs, reduced lead times, increased capacity, easier tasks, and happier customers, all resulting in a more successful business. This point has been proven over and over again, first by Toyota in Japan, and then by many other businesses throughout the world.

The same philosophy goes for green. By developing a set of environmental wastes, you establish a list of criteria that are based on what the environment does not perceive as positive, good, or valuable. By checking these criteria against your value stream, you identify all the negative impacts on the environment that exist within your value stream and then follow a specific waste elimination process that results in reduced costs, increased value, and competitiveness.

The seven green wastes are as follows, along with symbols used to identify each:

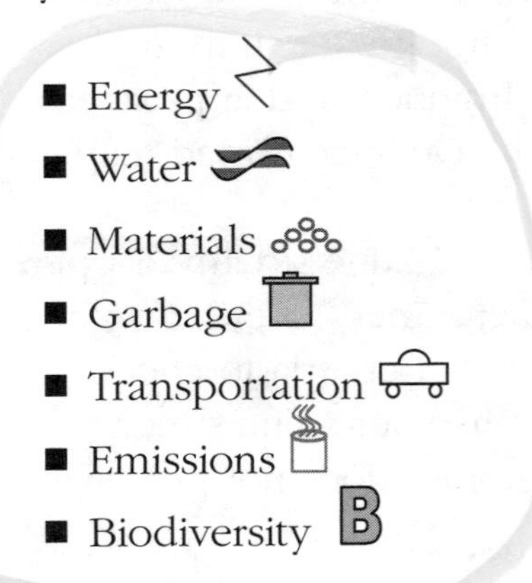

- Energy
- Water
- Materials
- Garbage
- Transportation
- Emissions
- Biodiversity

THE GLOBAL REPORTING INITIATIVE

The seven green wastes encompass of sustainability the environmental dimensions covered under the internationally accepted standard for sustainability reporting, the Global Reporting Initiative (GRI). The GRI is a not-for-profit, multistakeholder institution that provides a universally accepted reporting framework for organizations to disclose their sustainability performance. The initiative is governed by a board of directors, who utilize technical committees and organizational stakeholders in order to ensure the credibility, completeness, continuous improvement, and global application of the framework. The framework is used to benchmark, compare, and promote an organization's sustainability performance in a format that is credible, transparent, and confirms their commitment to sustainability. This framework is broken up into the three dimensions of sustainability: environmental, economic, and social. For sustainability reporting, the GRI is the gold standard, which is why it was used as the reference to ensure that the seven green wastes completely cover all of the potential environmental impacts a company may create.

The great thing about this approach to going green is that you can use green-value-stream, also known as Lean & Green, thinking as a standalone, continuous-improvement approach or easily integrate it with existing continuous-improvement efforts. For example, if you are currently engaged in a lean initiative, you know that the fundamental purpose of lean is to eliminate waste. However, traditional lean thinking — although providing some environmental benefit — does not specifically focus on environmental impact as a waste. By adding environmental impact as an eighth waste (broken down into the seven subwastes listed in the following sections), you can use existing lean teams, resources, and infrastructure to eliminate this specific environmental waste.

Chapters 3 through 9 discuss each of these green wastes in detail, but the rest of this chapter offers a short summary of each so that you can begin to identify the green waste in your company.

The First Green Waste: Energy

The waste of energy comes from the fact that you are paying to consume more energy than required from a source that negatively impacts the environment, such as coal. Instead, you could be using only the energy required, and then move toward getting it from a source that is clean, green, and self-produced—in other words, getting that energy for free from the sun, wind, or other renewable sources. Such self-sustaining renewable energy eliminates both the economic and environmental wastes of energy.

Minimizing the overuse of power from things such as lighting, motors, and electronic equipment means that you are conserving a great deal of energy—energy that would need to be paid for—resulting in dramatic financial savings, not to mention the environmental benefits. These savings can be easily achieved through the use of low- or no-cost conservation techniques or through the use of energy-efficient products, both of which have rapid payback periods. Providing your own green power has also become attractive, and with the soaring cost of energy, many businesses (including Google) are jumping on this opportunity. Solar power has also come a long way in the last few years and is now providing payback in as little as five years in some areas. Today's solar panels have a lifespan of more than 25 years, which means that you enjoy the economic and environmental benefits for a number of years to come.

The Second Green Waste: Water

Water waste comes from paying to use more water than you need and paying again to have it taken away and cleaned. Instead, you could use only the water you need, move toward getting water for free by harvesting rainwater and finally moving towards the continual reuse of water. You can easily minimize water consumption by replacing water-consuming fixtures and tools, such as toilets, sinks, washers, and sprayers, with more water-efficient ones. Making these changes is easy and inexpensive, providing you with years of savings.

Harvesting rainwater to use for irrigation, for the flushing of toilets, or for any process activities saves you the cost of having to pay for that water. Reusing water by cleaning it yourself is the end goal in eliminating both the economic and environmental wastes of water

The Third Green Waste: Materials

Material waste comes from a global design flaw—designing virgin raw materials into products that are designed to end up in the landfill. Instead, you could move away from this linear "cradle-to-grave" approach to a cyclical cradle-to-cradle approach of designing products to come back to you so you can reuse the materials in them to make another product. This concept, by William McDonough and Michael Braungart, is the end goal in eliminating material waste.

When you design a product that can come back to you so that you can make another product out of it, the material costs of making that second product are dramatically reduced. As you might imagine, the savings are astronomical, which is why many companies are already putting this practice into place. Interface floors, under the guidance of Ray Anderson, and with the support of passionate engineers such as Dave Gustashaw, eliminates this waste elegantly by "renting" carpets to customers. When the carpet wears out or gets dated, it can be ripped out, broken down, and turned into a brand new carpet. This process can be repeated indefinitely.

The positive environmental impact aside, the financial impact is dramatic from reduced material costs because the company is now getting its material much cheaper or for free, they establish long-term repeat customers, plus they attract new customers because of the lower total cost of ownership.

The Fourth Green Waste: Garbage

Garbage waste comes from paying for something that you will throw away, something that has caused negative environmental impact to produce, and then paying again to have it disposed of. By minimizing and moving toward eliminating the amount of items you throw "away" (there is no "away"), you save on both ends by not having to pay for it in the first place and by not having to pay to have it taken away. Packaging and scrap offer a great place to start when eliminating garbage waste.

By choosing recyclable, reusable, or biodegradable packaging materials for incoming goods, you save on the disposal costs associated with those items as it is cheaper to recycle than to dump in a landfill. Using biodegradable, recyclable, or reusable packaging for outgoing goods not only offers cost savings to your customers, but also provides great marketing potential, helping to attract new customers and employees as well as retain current ones. Of course the end goal is to not create any garbage at all but rather create value or food for something else.

The Fifth Green Waste: Transportation

Transportation waste comes from paying for excess and unnecessary travel that results in negative impacts on the environment from the burning of fossil fuels. Instead, you can use only the amount of transportation that is required and then move toward using environmentally friendlier modes of transportation (such as hybrid or electric vehicles charged by renewables) in order to eliminate this waste. Not only do you gain the environmental benefit of burning less fossil fuels, but you also save by not having to pay for those fossil fuels—the cost of which has been rising precipitously and will continue to do so in the future.

Sources of travel and transportation waste are rampant throughout organizations today. Everything from shipping materials and products over long distances (instead of buying or producing locally), to unnecessary business travel that could be avoided by using existing technology, to sending documents and items via courier that could be sent electronically or could be consolidated—these are all wastes, and minimizing them will save both money and the environment.

The Sixth Green Waste: Emissions

Emissions waste comes from paying to create and discharge pollutants on-site, and then being subject to the fines and levies associated with doing this. Instead, you could work to minimize the creation of emissions on-site and start to save on the fees, fines, and levies that come from discharging emissions. Then you could work to eliminate this waste by looking at the source of the pollutant or contaminant and beginning to replace these with ones that are less harmful or nonharmful. Many companies are replacing toxic solvents, varnishes, and epoxies used in production processes with ones that are less toxic, resulting in savings on both the cost of the substance and the costs associated with putting the harmful substance into the air or ground.

The Seventh Green Waste: Biodiversity

Biodiversity waste comes in two forms: first from the direct destruction of flora, fauna and organisms resulting from the building of infrastructure, secondly from the overharvesting of natural resources. The waste of biodiversity in its first form comes from the fact that you are paying to destroy biodiversity such as trees, landscapes, and watersheds so that you can build infrastructure, and then paying to replace the services that these natural elements could have provided for free. Instead you could build in such a way as to minimize or eliminate the destruction of biodiversity or, if already destroyed, work to regenerate it. A number of organizations have done this, putting up green roofs or planting more vegetation that provides environmental benefits and, in turn, saving money from storm-water management costs and heating and cooling costs, to name a few.

The waste of biodiversity in its second form comes from overharvesting resources faster than they can regenerate themselves. This results in the loss of potential future income. Instead, you could harvest at a rate that allows you to sustain harvesting for a much longer time, providing more income over the long run.

In Section II, you will find step-by-step processes for identifying, measuring, minimizing, and eventually eliminating each of the seven green wastes. But first, we will explore the concept of the green value stream.

Chapter 2

Understanding Your Green Value Streams

The idea of the value stream is not a new concept, nor is it a complicated one. However, in order to understand a green value stream (GVS), you need a basic understanding of a traditional value stream, an idea that comes from lean manufacturing.

So what is a value stream? A value stream is the flow of *all* the activities that your organization performs in order to provide a product or service to your customer, regardless of whether the activities are considered value-adding or non-value-adding. This includes the production activities required to change the form and function of something into a usable product or service.

In addition, you also have administrative functions that support the production activities, such as building operations, accounting, human resources, customer service, and so on. By following the lean approach, you can systematically eliminate waste and increase the value in your value stream.

Note that a business may have many different value streams within its operations. For example, if a company produces two different products that go through different processes and need different administration and support, you have two different value streams.

Starting with an Example: Value Streams versus Green Value Streams

To give you an example of a value stream, I will refer to a fictional company, "Greanco," that serves throughout the book.

Greanco is a manufacturer of metal furniture to the corporate and educational markets. Each product they make must go through the same processes. Greanco takes raw materials such as steel and, through a series of different processes or activities, changes the form and function of these raw materials to create a finished product that they then sell to their customers. In order for Greanco to create a finished product, the company performs the following activities:

- **Receiving:** Receiving of raw materials
- **Forming:** Taking sheets of steel and bending, cutting, and shaping them to form pieces that make up the frame for the furniture
- **Welding:** Welding the different formed pieces together to complete the frame of the furniture
- **Washing:** Cleaning the metal frame of any debris/grease that results from the first two activities
- **Preassembly:** Attaching the metal hardware
- **Painting:** Powder coating the metal frame and hardware
- **Final assembly:** Attaching the wood and cushions required for the particular piece of furniture
- **Shipping:** Shipping finished goods to customers

Each of these activities is a "pure" value stream activity, because it changes the form and function of something. In addition to these pure value stream activities, Greanco requires the following administrative and business activities to support the pure value stream activities:

- **Administration:** Answering the phone, opening mail, staffing the reception desk, attending to travel details, planning meetings

- **Accounting:** Performing accounts payable and receivable functions, generating financial statements
- **Operations:** Purchasing, scheduling, environmental, health and safety
- **Human resources:** Hiring, firing, administering benefits
- **Sales and marketing:** Performing all sales and marketing tasks for the company
- **Engineering:** Designing, supporting, and developing products
- **Building envelope:** Maintaining the building through heating, ventilation, and air conditioning (HVAC) activities, lighting, plumbing

These activities are *not* pure value stream activities because they do not change the form and function of materials to produce a product or service, but they are necessary to support those pure value stream activities. Note: throughout the book these activities are referred to as "The Overall Building."

Now that you have seen a traditional value stream, a green value stream focuses on the environmental impacts, good or bad, associated with your value stream activities. Here are a few examples from Greanco (this list is by no means exhaustive):

- **Receiving:** The pollution resulting from the transport of raw materials to the factory
- **Forming:** The energy used to run the forming machine
- **Welding:** The energy used to run the welding machine
- **Washing:** The water used and contaminated from the washing process
- **Preassembly:** The garbage generated from the packaging of the parts
- **Painting:** The energy used to run the oven and emissions from paint fumes
- **Final assembly:** The energy used to run the compressor that feeds the tools for assembly
- **Shipping:** The pollution resulting from the transportation of the goods and the garbage from packaging material.

Similarly, here are a few examples of the environmental impacts of the administrative and business activities that support the pure value stream activities:

- **Administration:** The energy used to run computers and printers
- **Accounting:** Environmental impact resulting from making the paper and envelopes used to cut and send paper checks.
- **Operations:** The energy used to run all the computers required to manage production
- **Human resources:** The paper, other materials, and energy used to administer payroll and benefits on a computer system
- **Sales and marketing:** The emissions from the transportation required to make sales calls and energy used in computers
- **Engineering:** The energy used to develop and test the products and garbage resulting from failed products
- **Building envelope:** Energy to heat and light the building plus the water used in plumbing and irrigation

As you can see, you can identify a green value stream by associating environmental impacts, good or bad, with each of the activities in your value stream. The trick is being able to identify all the environmental impacts and opportunities in a systematic fashion, and then being able to systematically eliminate them. When you identify the seven green wastes (summarized in Chapter 1 and discussed in detail individually in Chapters 3 through 9, respectively), and move towards their minimization and eventual elimination, you systematically increase the "green" in your green value stream, and this is the green value stream approach, also known as Lean & Green.

The Green Value Stream (GVS) Approach

The primary focus of a traditional value stream process is to eliminate the seven wastes associated with lean manufacturing. Companies put such a process into place in order to identify waste and then follow the procedure by which those wastes will be eliminated. Likewise, a green value stream (GVS) process seeks to eliminate the seven green wastes and establish a procedure for eliminating those wastes.

By following the overall GVS process, you can systematically eliminate the environmental wastes in your organization and move toward a greener way of being, allowing your organization to achieve dramatic results for the environment and the bottom line. You can tackle the wastes one at a time, or you can do them together; just remember, if you're not looking for any of the seven potential green wastes, you're losing out on a huge opportunity and throwing money down the drain.

What follows is the overall GVS process that can be used by any organization wanting to go green. After reading this book, you will understand the step-by-step waste-elimination process for each of the seven green wastes and be able to follow and execute each one of the steps in the overall GVS process. This enables you to systematically move your organization toward environmental sustainability and realize the benefits associated with going green.

The steps in the overall GVS process are:

1. Gain management support, develop a vision, and appoint green champions within the organization
2. Shift your thinking; grasp the concept of looking at things from the perspective of the environment
3. Be able to list each of the seven green wastes and the step-by-step process for eliminating them
4. Create a current-state green stream map by identifying and measuring the seven green wastes in your value stream
5. Create a future-state green stream map by following the minimization step for each of the green wastes; then implement the solutions on your future-state map continuously until you have minimized all wastes as much as possible
6. Pursue the perfection of total elimination of all seven wastes by following the final steps in the elimination of each of the green wastes, thus allowing you to achieve your green state
7. Carry GVS to your supply chain

This process, applicable across most, if not all, businesses or organizations, allows you to move systematically toward increased harmony between business and the environment.

Armed with an understanding of the overall green value stream process, and having covered a major step in the process (the shift of thinking), you are ready to look at green stream mapping, which will visually illustrate the GVS process.

Green Stream Mapping

Green stream mapping is a green spin on the lean tool of value stream mapping. Value stream mapping involves drawing a map of the value stream activities and the overall building activities that are performed in creating a product or service, and then identifying the lean wastes associated with those activities. By mapping, you visually illustrate the value stream activities and the opportunities for improvement in a *current-state* value stream map. Once you come up with solutions for eliminating the waste, you create a *future-state* value stream map illustrating what your value stream will look like when you implement the solutions.

Green stream mapping uses the same concept as value stream mapping, except that instead of illustrating lean wastes, you are illustrating the green wastes to develop a current-state green stream map (GSM). You then develop solutions for minimizing those green wastes, and create your future-state map. Finally, you pursue perfection (your green state) by developing solutions to eliminate the wastes completely. Figure 2.1 shows an example of a traditional current-state value stream map. Figure 2.2 is a current-state green stream map. Note that everything outside the value stream itself (the "Overall Building" activities) is denoted on the GSM as "Admin and Support."

Creating Your Current-State Green Stream Map

There are three parts to creating a current-state green stream map. The first part involves mapping the value stream; the second part involves observing and identifying each step or process; and the third part involves identifying and measuring the seven green wastes in your value stream activities, and then illustrating those on the green stream map.

Map What You Do

We'll start in the logical place: the first step. If you have already been practicing lean, chances are you've already done this first step in the form of a value stream map. If so, take your value stream map and remove all the lean symbols from it. If you do not already have this value stream map, it is fairly simple to map what you currently do.

As already mentioned, a value stream encompasses all the activities that you do in order to get a specific product out the door to a customer.

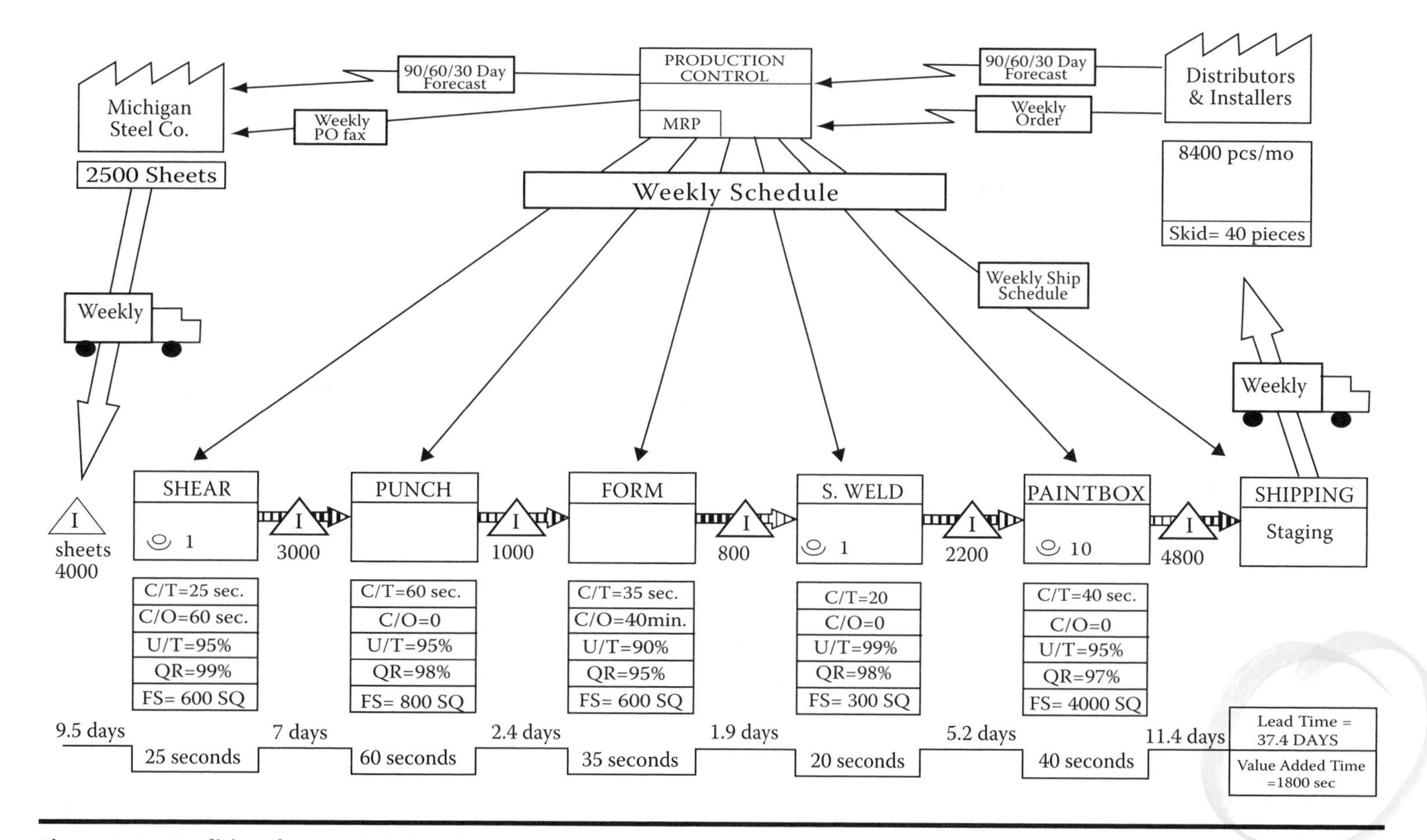

Figure 2.1 Traditional current-state value stream map.

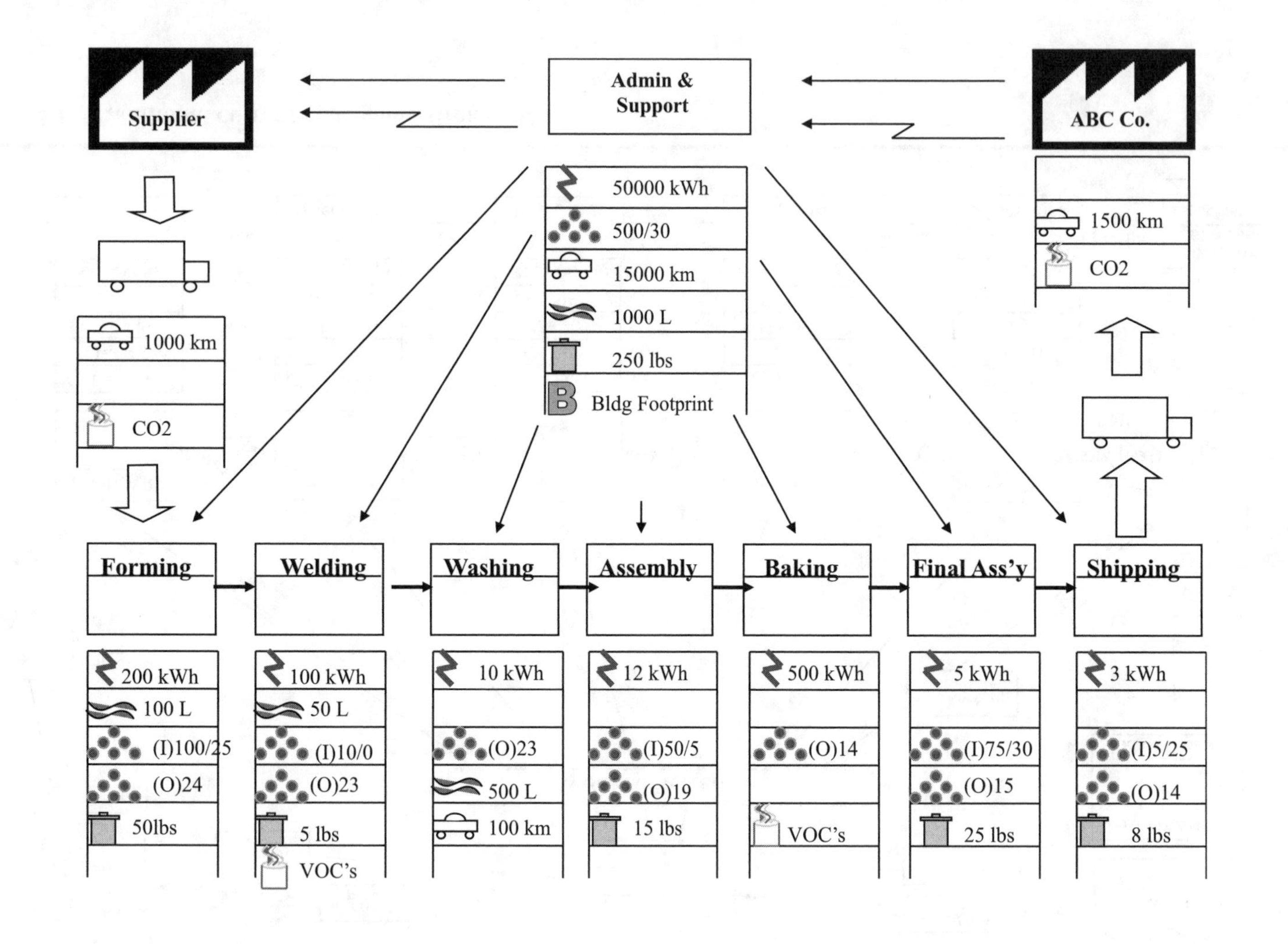

Figure 2.2 Current-state green stream map.

Figure 2.3 Mapping the customer and supplier functions

Grab a sheet of paper (100 percent recycled paper is best) and go out to the floor where the product is being built. You may want to take the product/production/operations manager with you or someone, who knows the steps that it takes to get the product out the door. You are now ready to start mapping your current state. As an example, I use Greanco to help work through the process.

Start the map by drawing the customers and suppliers (include suppliers of both value stream activities and the overall building), using the factory icon in the top right- and left-hand corners, as shown in Figure 2.3.

Identify Each Step or Process

Next, identify each of the steps or processes required to get your product or service out the door; these are the pure value stream activities that change the form or function of something so that it has value to your customer. Try to focus on the major blocks of work and not the individual operations within a larger process. For example, welding has many steps, such as turning on the welding machine, clamping the metal together, heating the metal, and so on. Focusing on the larger blocks of work prevents the map from becoming overwhelming and crowded.

Greanco makes metal furniture, so the major steps or processes are:

- Forming
- Welding
- Preassembly

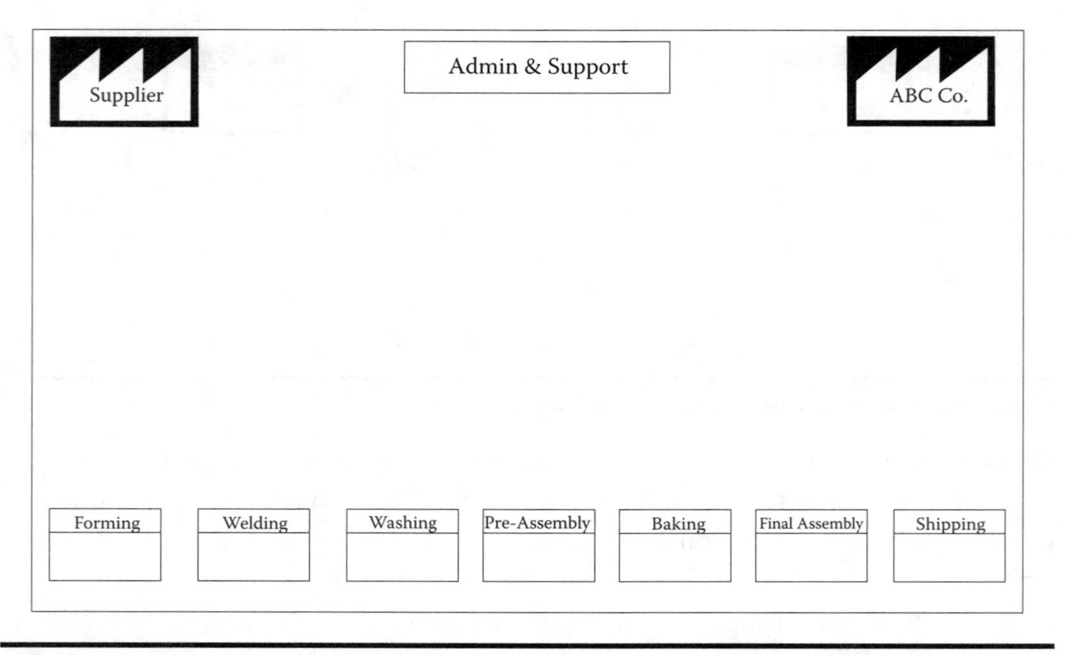

Figure 2.4 Mapping the administrative function and the value stream activities.

- Washing
- Painting and baking
- Final assembly
- Shipping

You also want to identify the administrative functions required to support the entire value stream, the "overall building" activities. It is important to note all these activities, as Greanco does, but for mapping purposes, group them all together as Admin and Support. Illustrate the overall building activities by using a process box and labeling it as "Admin & Support," as shown in Figure 2.4.

At this point, you have mapped the processes involved, taken into consideration the support functions of the value stream as well as the operation of the building itself, and identified your customers and suppliers.

Map How You Receive and Send Materials and Information

Next, illustrate how you receive and send materials and information. Illustrating that you receive and send materials is done using the truck symbol, along with an arrow to indicate the direction in which they are flowing. Indicate information flow using a broken arrow for electronic and a solid arrow for manual or paper flow. Last, show the flow or movement of a product or service through the processes required to create the final product.

An example of material and information flow is illustrated on Greanco's green stream map in Figure 2.5.

Draw in Data Boxes

Next, draw in data boxes, which provide a place to show the metrics used to quantify the green wastes. See Figure 2.6. There are a couple points to note here. First, there should be at least five or six lines in the data boxes to provide enough room to insert the symbols and metrics for the green wastes. Also, draw a data box under the Admin and Support box; this is the place to insert the waste symbols and metrics for overall building activities. Draw a data box under the customer's factory icon as well; this is where to indicate the emissions and any other green wastes associated with the customer's use of the product or service. You also want to draw a data box under the truck symbol to represent the receiving of goods, as this will account for a large part of the travel and transportation waste.

Identify the Green Wastes

At this point on your green stream map, you have illustrated your value stream—all the major activities you perform in the production of a product or service. Now, move on to the second part of creating the current-state green stream map, looking at each activity or process from the perspective of the environment. To accomplish this, review the identify step for each of seven green wastes, and then identify which ones are present in each activity, and indicate the presence of each waste by drawing its symbol in the appropriate data boxes under each activity in the green stream map.

A few notes here: For all processes that have materials flowing through them, even if the process itself does not add any materials, place a materials symbol. You do this to capture any impacts or changes on the material

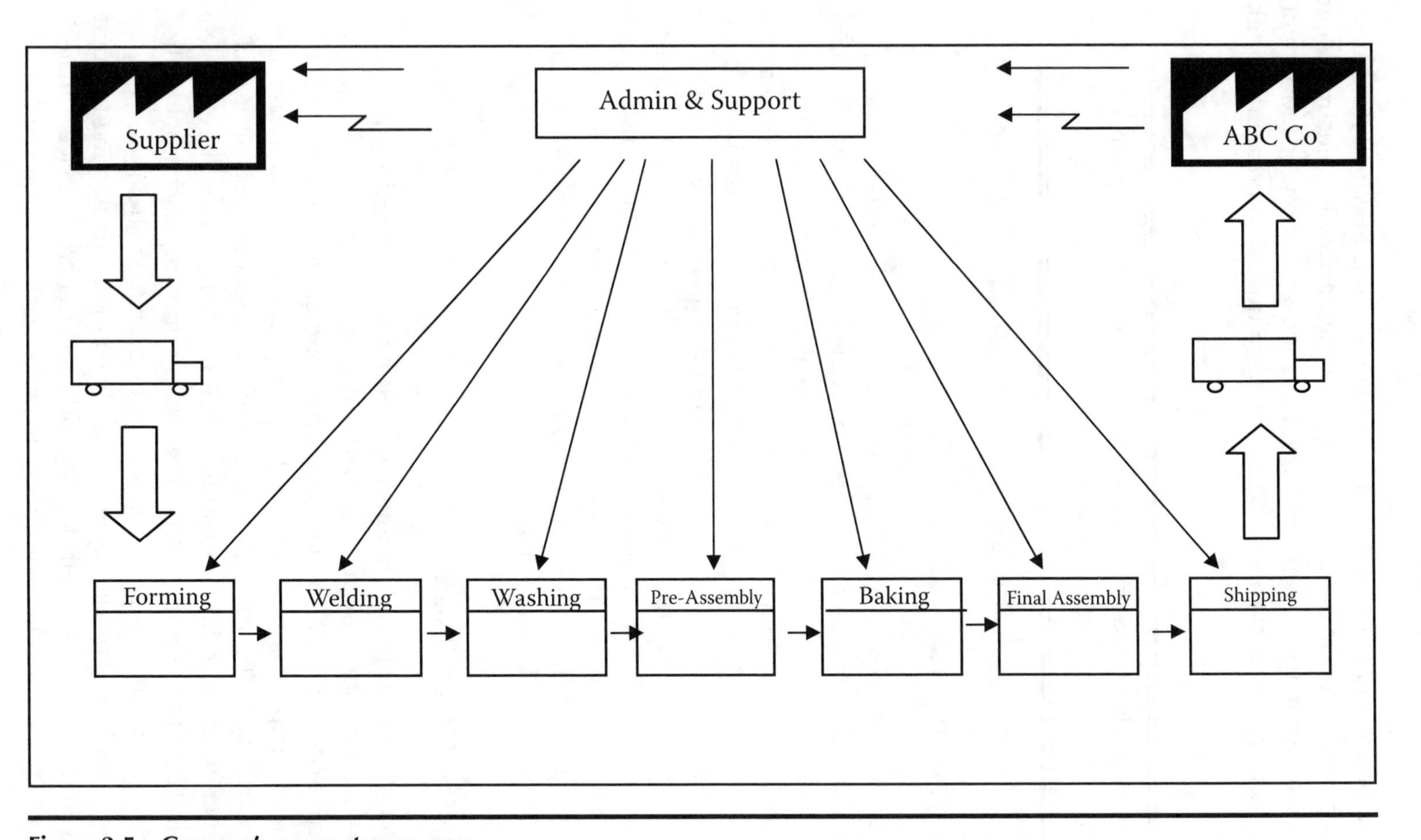

Figure 2.5 Greanco's green stream map.

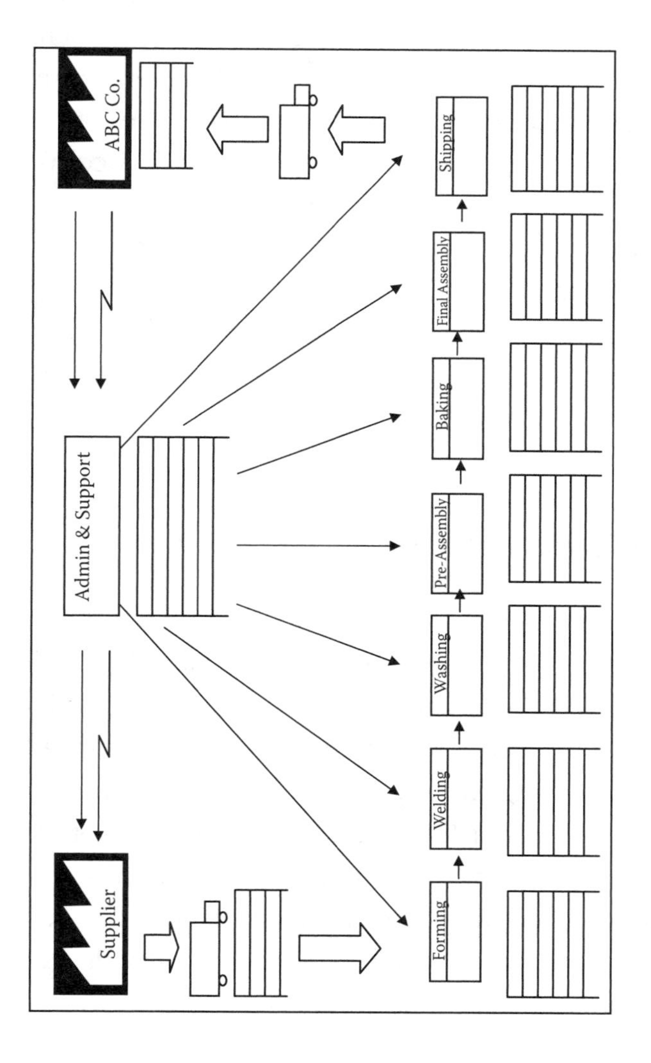

Figure 2.6 Green Stream Map before inserting current-state metrics.

outputs, denoted by (O), due to the process itself. For processes that have extra materials being added as inputs, insert a second materials symbol to capture this information, denoted by (I). An example of what your map should look like after this step is shown in Figure 2.7.

Measure the Green Wastes

Your current-state green stream map now shows all the activities you perform and identifies the negative environmental impacts, but does not show how much. In the third and final step in creating your current-state green stream map, you will measure the amount of the green wastes. To do this, you follow the Measure step for each of the green wastes, as detailed in Chapters 3 through 9. Then, sum up the total amount of each waste (total amount of energy, total amount of water, and so on) for each value stream activity and for administrative and support functions, and insert the total beside the appropriate symbol under each activity. An example of a completed current-state green stream map is shown in Figure 2.8.

What you have now created is a current-state green stream map that visually illustrates what and how much of the green wastes are present in your organization's value stream. Another way to look at it is that this current-state green stream map provides you with a visual image of green opportunities.

Creating a Future-State Green Stream Map

To develop your future-state green stream map, you need to follow the Minimization step for each of the green wastes is shown in Part II in order to develop solutions for minimizing these wastes. Once you have a solution to minimize one or a number of the value stream activities for the overall building, illustrate that fact by inserting the symbol for the waste you are minimizing inside the appropriate header box. Beside the symbol, note the reduction that will be achieved. To complete your future-state map, adjust the values in your data boxes to reflect the improvements. To do this, simply subtract the amount of waste reduction your solution will provide from the total amount you entered on your current-state map for that waste.

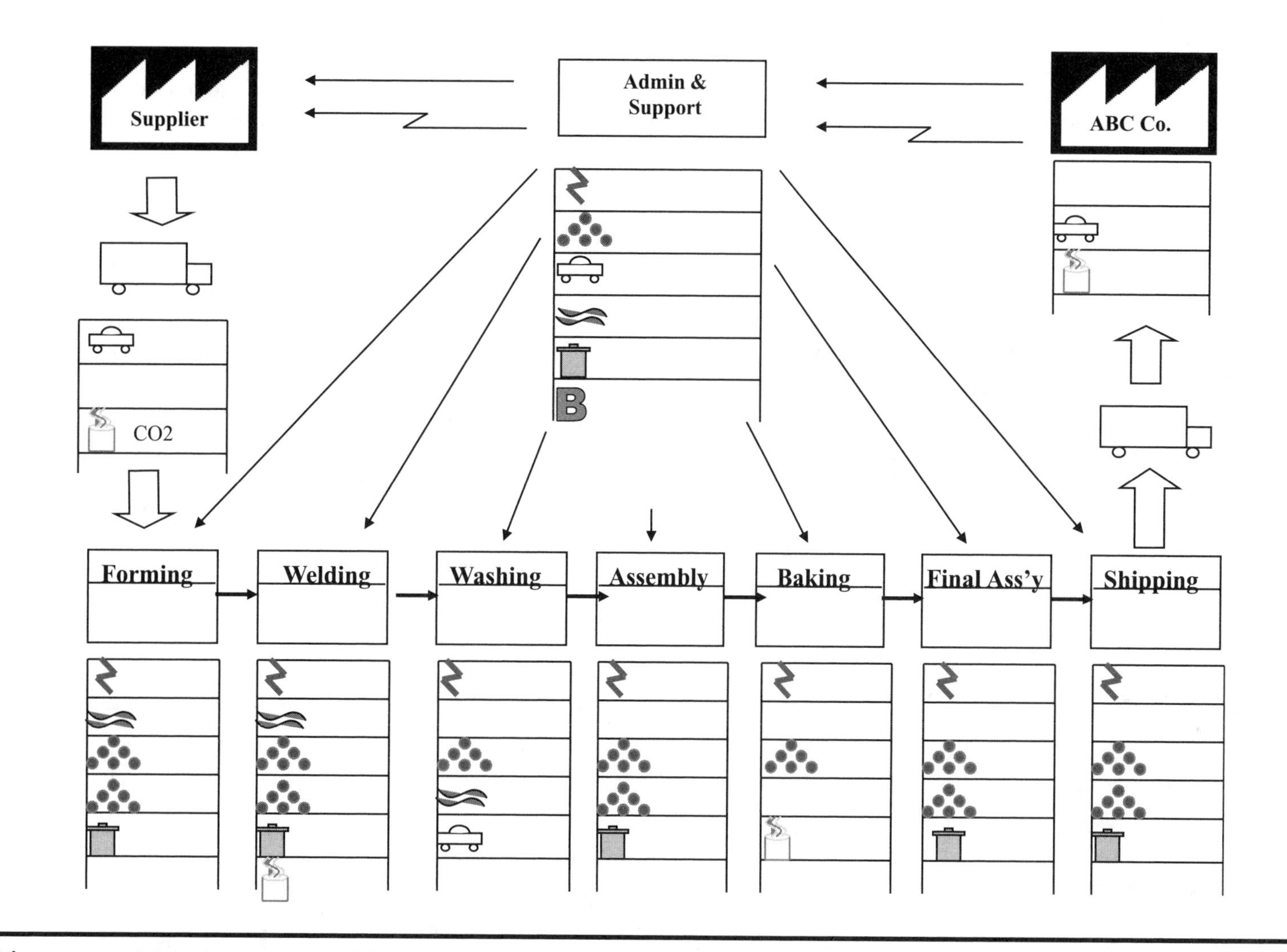

Figure 2.7 Green Stream Map with green wastes identified

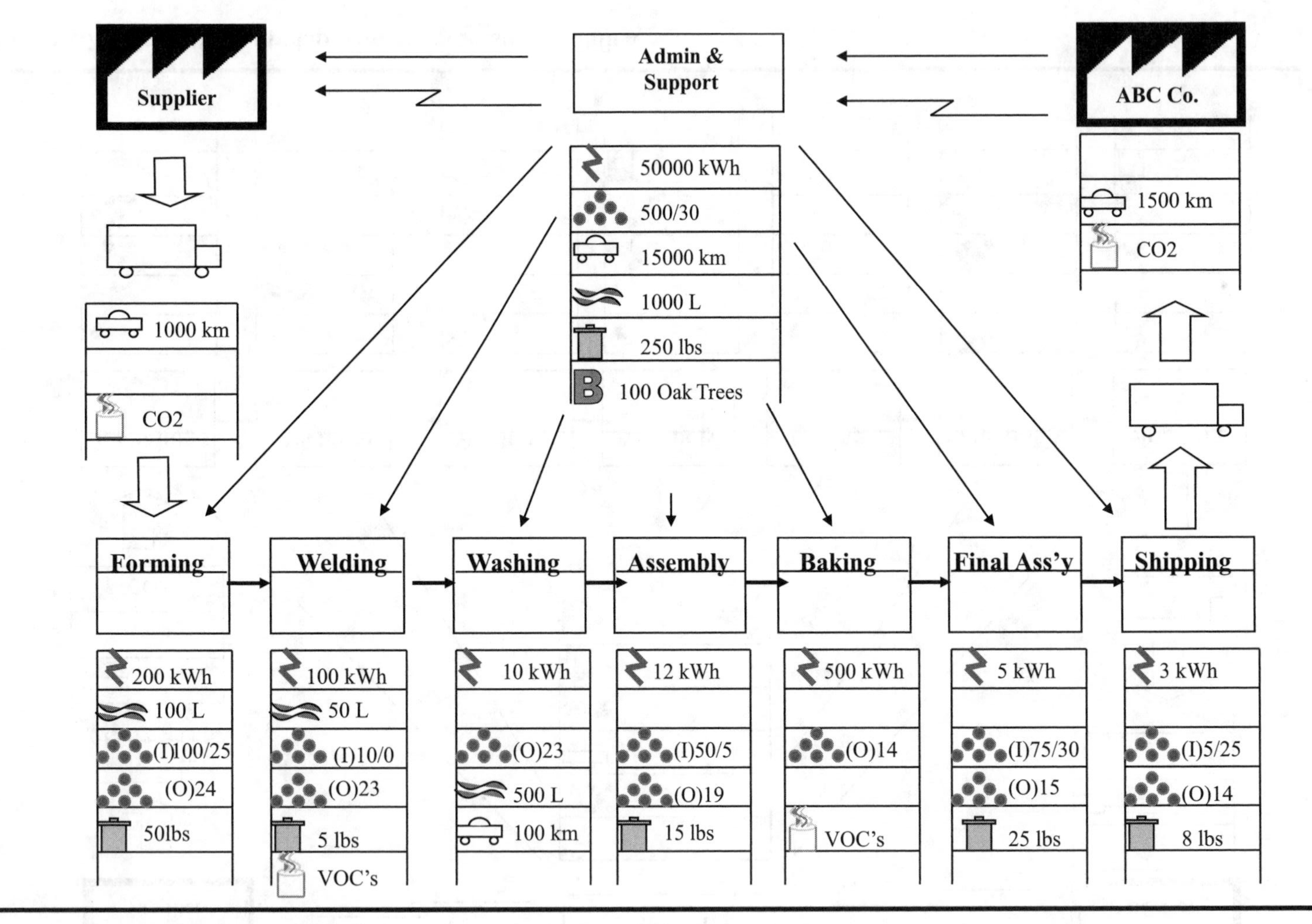

Figure 2.8 Example of a completed current-state green stream map.

Greanco has identified a number of solutions on its future-state green stream map, and Figure 2.9 gives you an idea of what a future-state green stream map looks like.

Please note that once you implement the solutions on your future-state map, it is no longer your future-state map; instead, it becomes your current-state map because it illustrates your current situation. By going back and developing solutions to further minimize your green wastes, you create another future-state map and so on. You want to repeat this process over and over again, until you have done all that you can to minimize all your green wastes.

Achieving a Green State: The Pursuit of Perfection

Once you have minimized all your green wastes as much as possible, you are ready to pursue perfection: eliminating the waste completely. Once you have the solutions for eliminating that waste, illustrate on your map that you have a solution by inserting the symbol of the waste you are eliminating inside of the appropriate header box. Next to the symbol, note the result of the solution. To complete your green-state map, enter the result of the solution in your data box values by changing them to reflect the elimination efforts or removing the eliminated waste completely.

This version of your future-state map is known as your "green-state map," because implementing those solutions will put you into your green state. The difference between your future state and your green state is that, in your future state, you focus on minimizing, while in your green state you focus on eliminating. Calling it your green-state map implies that you have minimized all the wastes as much as possible and are moving on to the next step of eliminating the wastes to achieve your green state.

Figure 2.10 illustrates what a green-state map looks like for Greanco.

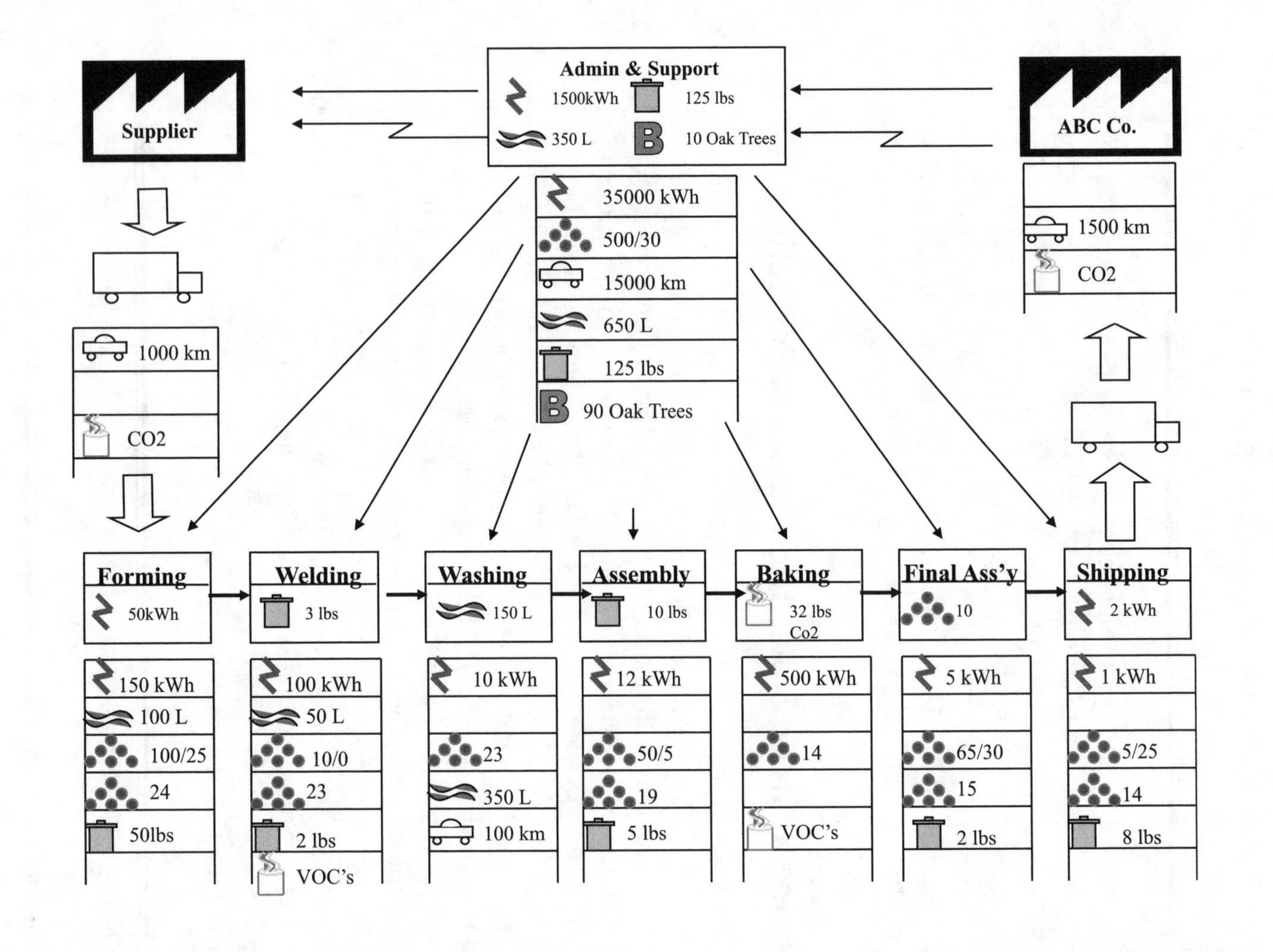

Figure 2.9 Completed Current-State Green Stream Map with green wastes identified and measured.

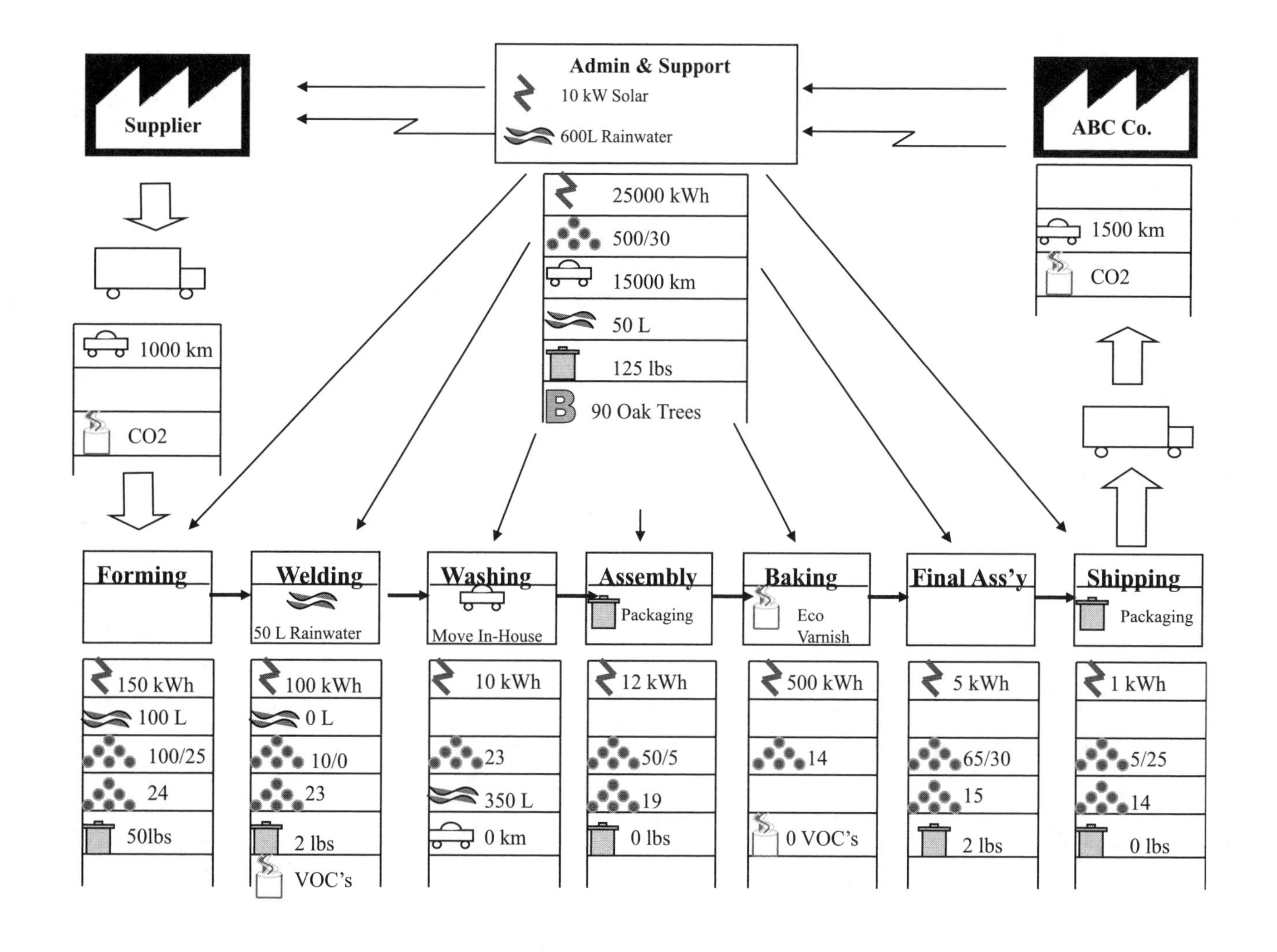

Figure 2.10 Greanco's green-state map.

Gaining Corporate Support for the GVS Process

With knowledge of the overall green value stream (GVS) or Lean & Green process, and having seen it visually illustrated in a green stream map, the question then becomes how to get the process started in your organization. The GVS approach, like lean, needs to start with a commitment from top management to go green, followed by a vision of what it will be like when you get there. Top management commitment and and a company vision will help to create a green culture. Without a culture that is continually moving toward an encompassing and agreed-upon vision, and that fosters the necessary values to support the vision, it is difficult to even start—never mind sustain—a green value stream approach. *With* this culture in place, led by a green champion and following the GVS process, green will become natural and will flow smoothly and easily.

Gaining a Commitment from Top Management

Commitment from top management to support the green efforts of the organization enables green change to happen. Commitment may come in many forms, from a simple verbal commitment to a signed charter or written letter. (Of course, it is always best to get a commitment in writing whenever possible.) In order to convince top management to give this commitment, show them the many examples of financial savings that you find throughout this book or go over the business code as discussed in the introduction to this book. You may also want to do a little research to see what your competitors are doing. If they are going green, use that information to demonstrate the savings and success they are getting. You can also set in motion a few pilot projects, using the approach in this book, to realize some immediate savings on some of the low-hanging fruit.

Setting Your Green Vision

A vision is the reason your organization exists, inspiring and guiding everything your organization does and giving meaning to the day-to-day activities. A vision is not meant to be attainable, but rather a picture of what your organization considers perfection.

As you may expect, a green vision is a picture of the ideal environmental impact you want your organization to have. Everything you do from the time you establish that vision will be in pursuit of achieving this ideal environmental impact.

The green vision differs from organization to organization. For those who make products or offer services that help others improve their environmental impact, a vision may be to put those products and services into the hands of the most people possible. For others, it may be to neutralize the environmental impact of their internal operations. Others may include their supply chain. Still others may combine all these and strive to have an overall positive impact. An example of a green vision is:

> "Greanco is committed to environmental sustainability. With this commitment, Greanco strives to have an enterprise-wide positive impact on the environment. Our enterprise includes our supply chain, internal operations, and end users of our product."

Set your green vision as high as possible, to include as much of your enterprise as possible, and to have an overall positive impact on the environment. Aim for a result that would mean the planet would be better off environmentally because your organization is not in business. At a minimum, your green vision should be to have an environmentally neutral impact from internal operations. Your green mission will then be the general way in which you are going to achieve your green vision, such as your approach for selling more products that help customers mitigate their environmental impact, or the approach for the greening of your operations.

Establishing Your Green Champion

A green champion (or group of green champions known as a green team) will provide leadership for the green process. This person should be someone who is already passionate about sustainability and green issues, who can motivate and inspire others to adopt green practices, and who is an overall leader. The champion needs to have a fairly good understanding of green, the end goal of sustainability, and at least a basic knowledge of the tools and techniques used to go green. If you can't find someone that already has these qualifications on his or her resume, choose an environmentally passionate person and train that person on the missing attributes.

A green champion's mandate is to drive environmental initiatives that move your organization closer to its environmental vision and that foster a green culture. For this reason, the champion needs support from senior management. If the green champion has other full-time duties, blocks of time need to be cleared during which the champion can focus solely on green initiatives. You might pass some of the champion's duties to others or hire others to relieve some of those other duties. Support also comes in the financial form of providing a budget for the champion to work with, providing extra training, and offering incentives for completing projects or achieving goals.

To prevent the green champion or team from becoming siloed from the rest of the organization, senior management should allow anyone associated with the green process to engage other employees. A title such as Sustainability Coordinator or Green Champion will go a long way toward reinforcing this individual's role and encouraging cooperation with his or her efforts. With support in place, it is then up to the green champion to initiate, lead, develop, manage, and see through to completion the environmental projects that will move your organization closer to its environmental vision. This person will also celebrate the little successes and report the results to senior management.

Establishing a Green Team

You may choose to enlist a group of green champions, known as a green team. The members of your green team should have the same qualities as the green champion described in the preceding section. However, make sure that each member brings a unique skill set to the team. Just as with a sports team, you want members who contribute different value, so choose people from different areas of the organization who have a passion for the environment—a person from accounting, a person from operations, a person from engineering, a person from production, a person from sales and marketing, and so on. Remember that, as with the green champion, you are looking for people who are passionate and motivated to improve environmental impact. With this passion and motivation, you can easily train them on the GVS process, so that they will have the ability to carry out the waste-elimination steps.

Next, you will need to decide how the green team will approach waste elimination. Will the team look at all the wastes in one activity, and then move on to the next activity? Or will the team look at one waste in all activities, and then move on to the next waste? How will it be decided what projects to work on first? (The largest waste generators that bring the biggest payback should be at the top of the list.)

With the team formed and a plan of attack in place, you need to assign roles and responsibilities for each team member. Who is going to do what? Will the whole team work together through each of the steps in waste elimination, or will team members be pulled in at different steps? Typically, it works best to have the whole team identify and measure, so that everyone has an understanding of the wastes that are present. Then it may make sense to break out the team—to have the production and engineering members develop solutions, for example, and have the accounting people work on calculations and prepare budgets for the projects. Also, consider who will be responsible for ensuring that solutions are implemented. Who is responsible for reporting to senior management?

To get the most out of your green team, communication among team members will be essential. Will the team hold regularly scheduled meetings or conference calls? Also essential to getting the most out of the green team is a reward system based on achieving the economic and environmental benefits. This may be the responsibility of the team leader.

With the green team in place, you are now ready to quickly and systematically realize the many benefits of going green.

The Bottom-Up Approach

Sometimes the top-down approach can be difficult to execute, especially in larger organizations. For this reason, you may choose to follow the bottom-up approach, whereby you follow this process in a single division or in an area over which you personally have control or influence. To follow a bottom-up approach, you only need to get support from your immediate superiors, but not from top management. Instead of creating a vision for an entire organization, you can simply create a vision specific to the area you are focusing on. With support from your immediate supervisor and a vision of what things will look like in your area, you can become the green champion for this area and use the GVS approach to realize immediate results. People will take notice and want to move green improvements to other areas of the organization.

The green value stream approach allows you personally to drive enormous positive impact on the environment. Although we can all do our part by being greener in our personal lives, and most of us do try to do this,

UNDERSTANDING THE TREE CONCEPT

A tree effortlessly continues to grow and develop year after year, using the power of the sun as its energy, with no negative impact on the environment. It does this all while providing a service (cleaning the air) and producing a product (leaves that, once they reach the end of their lives, are used as a nutrient to help the tree grow; some trees also provide fruits or nuts).

Now imagine a business that operates under the same premise, one that uses the free power of the sun or wind or heat of the earth to produce a useable product that, when it reaches the end of its life, could be used as a nutrient either to build new products or to be put back into the earth. In the case of a service company, imagine a service that has a positive effect on the environment.

Think about the implications of that vision. The kicker here is that this can be and is being done. Many of the ideas and concepts used in GVS, along with many of the solutions you will find as you go through the GVS process, already exist and are delivering real results to companies all over the world. GVS pulls it all together in a systematic and proven process to achieve quick and dramatic results easily. All that is required to go down this road is to shift thinking and paradigms from the current ways of operating and then to operate in a way that is aligned with the natural world.

we can have an exponentially larger impact by being greener in our professional lives. Saving energy at home has a small impact, but saving energy at work has a larger impact because it is a larger amount of energy. Your green efforts at work will result not only in a larger positive environmental impact, but also, as you will see throughout this book, in larger cost savings and other benefits that will surely help you on your career path.

Life Cycle Analysis

When undertaking any “green” initiative, it is crucial to look at the total direct and indirect environmental impact that will result. This is known as a Life Cycle Analysis or Life Cycle Assessment (LCA). When it comes to “green,” this is a very important concept. The whole point of going green is to have less overall impact on the environment. When making a change or comparing a number of options, the only way to ensure that you will have less impact on the environment (that is, that you will be "greener") is to know the total environmental impact throughout the entire life cycle of a product, service, or project.

Life-cycle assessments can get very complicated, and the details are beyond the scope of this book. Nevertheless, it is useful to have a general understanding of how LCAs are done so that you have an idea of the questions to ask that will enable you to make the greener choice. Essentially, a life-cycle assessment entails reviewing all of the environmental impacts that result from the use of a product, service, or undertaking a project. This includes environmental impacts from:

- Pulling raw materials out of the ground, such as depletion of nonrenewable resources and energy used to power the machines needed to obtain the materials
- Using energy to run machines or using chemicals in the manufacturing process
- Shipping finished goods to your customer or supplies to your door (the environmental impact of shipping supplies from overseas is much greater than supplies coming from a local supplier)
- Using the final product, such as burning gas in cars
- Disposing of a product when it reaches its end of life, such as "throwing away" plastics

All of these concerns are considered in a life-cycle assessment, and only when you take them all into consideration can you make an informed choice on what is the greener option. (In fact, you should also consider the social aspects, such as where something was made, under what conditions, and what labor was used. Since this book focuses on the environmental dimension of sustainability, we will stick to that area.)

The good news is that, most of the time, life-cycle analyses have already been done at a detailed level by a third-party agency, whether it be government, academic, or other. You can find many such resources, including a growing list of blogs that discuss these issues in detail, through a simple Google search. If such resources are not immediately available, you can use common sense and ask a few questions, such as:

- What materials are being used?
- How were they processed?

- Where did they come from?
- What impact does something have when in use?
- What about when it comes to its end of life?

Then you will have some basic information to compare your choices.

It is important to note that you should use caution when a supplier of a product makes green claims. Oftentimes, especially lately, companies exaggerate or completely misrepresent the environmental benefits or "greenness" of their product or service. This is known as "greenwashing," and you must avoid being duped by it. Being an informed consumer does not stop when it comes to knowing the facts about greener products; in fact, it requires extra diligence. Likewise, remember not to attempt greenwashing yourself, as it will have a tremendous negative impact on your reputation that will be very hard to recover from. Furthermore, there are watchdog organizations that can catch greenwashing and even make organizations subject to legal action for making false or misleading claims.

With these precautions in mind, let's move on to learning about the seven green wastes and how to root them out.

THE SEVEN GREEN WASTES

Chapter 3

The First Green Waste: Energy

Energy is often used to describe a broad range of activities, but in this context, as a green waste, it specifically refers to the consumption of electricity and fuels (such as natural gas) to power electrical and mechanical devices (such as electronics, heating devices, and machinery, as well as building items like lighting and HVAC systems). Other forms of energy, such as that used for transportation, are covered under the other six wastes discussed in Chapters 4 through 9.

Energy is a requirement for almost every aspect of a business, so much so that it would be hard to imagine running any part of a business without it. However, it is also one of the largest contributors to negative impacts on the environment. From a green perspective, therefore, energy represents one of the largest opportunities for improvement and savings.

Consider this: It is not necessarily the *use* of energy that is wasteful, but rather (1) the wasteful overuse of energy, (2) the dirty source of energy, and (3) the fact that you have to buy it from someone else. The ultimate goal of any green company is not only to switch completely to clean energy sources (such as solar or wind), but also to provide it for yourself. Imagine if you didn't have to pay for power because the sun or wind gave it to you for free. How's that for a competitive advantage?

Before taking on such a large endeavor, however, you first need to become more energy efficient with your current energy sources. If, through an energy-reduction initiative, you can minimize the amount of power you consume, you will minimize the amount of power you pay for, which not only helps the environment, but also provides quick savings and a healthier bottom line. Yet, companies struggle to reduce their energy consumption.

To reduce your energy wastes, first look at your value stream from thep-erspective of the environment and then follow the systematic energy waste-elimination process:

ENERGY WASTE ELIMINATION

Step 1: Identify the use and source of energy in each activity in your value stream and the overall building.
Step 2: Measure the energy used in each of these activities and the overall building.
Step 3: Minimize the use of energy in both your value stream activities and the building as a whole.
Step 4: Offset remaining energy requirements.
Step 5: Transition to the use of harvested renewable energy.

INTENDING TO USE SELF-GENERATED CLEAN ENERGY

Perhaps the most important step in the elimination of every waste starts with the *intent* to achieve the end goal, thus providing you with a target or destination to work toward. This intention provides meaning and value to each step in the process, because each step contributes to achieving the end goal. It's simple, but without such an intention, your organization may get lost or stray from the path.

Step 1: Identify the Use and Source of Energy in Each Activity

In this step, you are not measuring the amount of energy used (that comes in the next step), but instead are identifying all the activities that do consume energy, as well as the source of that energy. In order to identify what is consuming energy in your company, you need to walk out to the area in which the first activity of your value stream is located. You are looking for activities that consume energy, such as those that:

- Produce heat
- Produce light
- Create motion

More specifically, you are looking for electronic equipment, machinery, motors, heaters, ovens, lights, chargers, and so on. You can also see what is plugged into an electrical outlet or follow a gas pipe to see where it leads. Another great way to determine what is pulling energy is to ask the people who work in the area what they have to turn on or off. Finally, consulting the operations or plant manager should ensure that you have identified all the activities that are pulling energy. Remember that the point here is only to identify each item that is consuming energy, not to determine how much energy it is consuming.

Be sure to include activities powered by renewable energy sources, such as solar or wind power. You are not evaluating the "greenness" of the energy usage at this point, simply identifying the activities that consume energy.

When identifying the energy usage for the overall building, take the same approach by breaking down the areas of the building, such as the "Engineering Area" or the "Operations Area." This will allow you to focus your efforts so that you can be sure to identify all the activities that are consuming energy. Some typical energy consumers include computers and other IT equipment (such as servers), fax and copy machines, and buildingwide systems such as lighting and HVAC systems. (A facilities manager may also be able to help you identify HVAC items that consume energy.)

As you identify all the energy-consuming activities in your value stream and in the overall building, note them in the current-state section on the Energy Waste Elimination worksheet, as shown in Figure 3.1 and Figure 3.2. (Note that you should use a separate worksheet for each separate value stream activity or area of focus. Complete blank worksheets for each of the seven wastes are located in Appendix D.)

Once you identify the activities that are pulling energy, you then want to determine the source of the energy powering that activity. The "energy source" refers to what the electricity or fuel was derived from. For example, the electricity that comes into your building that is used to run your computers and other machines comes from a power plant somewhere. This power plant may produce electricity by burning coal, or it may be a nuclear power plant, or it may even be a solar- or wind-powered plant. Electricity will typically come from one source, such as coal, nuclear, or hydropower, but it may also come from a mix of sources. To determine the source, contact your electricity provider. This is the organization that sends you your electric or hydro bill every month. Call this company and ask about the source (coal, nuclear, and so on) of the energy being provided to you, and they will

Energy Waste Elimination Worksheet					
Activity or Area:		***Value Stream Activity 1: Forming***			
Current State					
Identify		**Measure**			
				Consumption	
Item	Source	Rate	Usage	Quantity	Cost
Forming machine	*Electricity*/gas*				
Crane	*Electricity**				
Computers	*Electricity**				
Workstation lights	*Electricity**				
* Electricity provided from a coal fired generating plant					

Figure 3.1 Energy waste elimination worksheet. Current state for value stream activity 1.

be able to tell you quickly. You can also do an Internet search for "energy source for [your area]" to determine the main source of power in your area. If power comes from more than one source, note each different source on the worksheet in Figure 3.1.

The same goes for fuel: The source could be natural gas, heating oil, and the like. In order to identify the source of the fuel, contact the fuel provider, that is, the gas or oil company that is sending you the monthly bill, and they will be able to tell you.

Repeat these steps for each activity in your value stream, as well as for the overall building. However, don't do more work than necessary. If you have established that your electricity is coming from a coal-fired power plant, all activities consuming electricity are from coal—no need to check each one.

To look for energy-consuming items, the Greanco green team went to the area of the first activity in the value stream. They first noted a machine that was forming the pieces of metal for the furniture, to which a pipe extended from the wall. The Greanco green team asked the operators what that pipe was. The workers said it was natural gas being fed to the machine to operate the heaters inside it. The rest of the machine was operating on electricity. The green team also noticed a crane loading the sheets of metal into the machine. There were computers in the area being used to tell the operators what needed to be made next. Finally, the team noted workstation lights on the ceiling, separate from the building lights. After they identified each item consuming energy, they needed to determine the source of that energy. They already knew that the fuel for heating the forming machine was natural gas. To determine the source of the electricity, they pulled out their utility bill, called the utility, and asked what the source of the energy was. The utility told them that it is a coal-fired generating plant. They then noted each of the items consuming energy—along with its source—on the worksheet in Figure 3.1.

When it came to the items that were consuming energy in the overall building, the Greanco green team decided to break this down into the different departments and areas of the business, so that they could ensure they had identified all items. They broke it down into the engineering, sales and marketing, accounting, and operations areas. In each area, the items consuming energy were similar and included computers, copiers, and faxes, as well as lighting and HVAC. The engineering department used similar items, but the green team also found some testing equipment unique to that area. Having identified the items consuming energy, they then needed to determine the source of that energy. In talking with the facilities manager, the green team found out that all energy being consumed was electrical except for the heating, which was from natural gas. They noted the equipment and their sources on the worksheet in Figure 3.2. (If you have a larger organization, you may also choose to note the location of the equipment, e.g., "engineering department" or "operations" on your worksheet.)

Step 2: Measure the Quantity of Energy Used

After you identify the use and source of energy, that energy needs to be measured. At first, determining the energy usage for each activity may seem difficult, but with a little practice it can be done easily and quickly. Because you now know all the items that are consuming power in each activity of your value stream(s) and in the overall building, you can look at each item and determine how much energy it is consuming. There are a few ways to go about measuring energy. Let's take a look at each in turn.

Using the Nameplate Method and Calculating

Every item that consumes energy is required to have a sticker or nameplate that tells you how much energy it consumes, typically listed as the amount of watts it pulls (like a 100-W lightbulb). If the item uses a fuel source of energy (such as gas or oil), the sticker or nameplate should list how much gas it uses in cubic feet or cubic meters or BTUs. For the sake of simplicity, we will focus on measuring electricity, but you can follow the same steps to determine your consumption of oil, natural gas, and other sources of fuel-based energy.

To measure how much energy you are using, start with the first item listed on your worksheets in Figure 3.1 and Figure 3.2. You first need to determine how much energy that equipment is consuming at any given moment. To do this, locate the nameplate on the device; it is usually located on the back, side, or the bottom of the device. What you are looking for is the amount of watts (W) or kilowatts (kW) (1 kilowatt = 1,000 watts) listed on the nameplate. If the nameplate doesn't list the watts or kilowatts, it may list the amps and volts. If this is the case, multiply the amps × volts to get the watts. For example:

$$2.5 \text{ amps} \times 20 \text{ volts} = 300 \text{ watts}$$

If the nameplate does not list these measures of energy consumption, it may show how much energy the equipment is consuming in some other units, such as horsepower (hp) for motors or British thermal units (BTUs) for

Energy Waste Elimination Worksheet					
Activity or Area:		***Overall Building***			
Current State					
Identify		**Measure**			
				Consumption	
Item	Source	Rate	Usage	Quantity	Cost
IT equipment	*Electricity**				
Lighting	*Electricity**				
Test equipment	*Electricity**				
HVAC	*Electricity*/gas*				
* Electricity provided from a coal fired generating plant					

Figure 3.2 Energy waste elimination worksheet. Current state: overall building.

heating and cooling devices. If this is the case, note that amount, such as 10 hp or 1,000 BTUs, and consult the following equations to convert to watts:

$$1 \text{ hp} = 746 \text{ W}$$

$$1 \text{ BTU} = 3.413 \text{ W}$$

The goal is to determine the amount of energy being used at any given time in the unit of watts or kilowatts, as this is usually how you are billed. If you can't find the nameplate, if it has fallen off, or if you can't get to it, note the device's model number and manufacturer and call them (its nameplate rating). They will be able to tell you how much energy it consumes at any given time. Once you have the amounts, note them all in the "Rate" column on your worksheet, as shown in Figure 3.3.

Now that you know how much energy the item is consuming at any given time in terms of watts or kilowatts, you need to determine how much of the time that item is consuming energy per day, week, month, or whatever period of time you choose. For example, if you have determined that a device consumes 100 W of energy at any given time, you now need to determine how often that device is used. Is it used 24 hours a day, 7 days a week? Eight hours a day, 5 days a week? For some equipment,

Energy Waste Elimination Worksheet					
Activity or Area:		***Value Stream Activity 1: Forming***			
Current State					
Identify		**Measure**			
				Consumption	
Item	Source	Rate	Usage	Quantity	Cost
Forming machine	*Electricity*/gas*	*17000W*	*16 hrs/day*	*5984 kWh/mth*	*$628.32*
Crane	*Electricity**	*6000W*	*7 hrs/day*	*924 kWh/mth*	*$97.02*
Computers	*Electricity**	*195W*	*24 hrs/day*	*103 kWh/mth*	*$10.82*
Workstation lights	*Electricity**	*1600W*	*16 hrs/day*	*563 kWh/mth*	*$59.12*
* Electricity provided from a coal fired generating plant					

Figure 3.3 Energy waste elimination worksheet. Current state: value stream activity 1.

usage time may vary, so try to get the average time used per day, week, or month. In order to determine the usage, ask the workers in the area how much they use it, and/or ask the production supervisor, operations manager, or plant manager to help you determine how much the equipment is used. If they don't know, ask workers to track usage for a day, week, or month. You can also observe and time the usage yourself. Once you have this information, note it in the "Usage" column on your worksheet, as shown in Figure 3.3.

Now you need to calculate the amount of kilowatt-hours each piece of equipment consumes per day, week, month, or year. To do this, multiply the amount of watts or kilowatts the equipment consumes at any given time by the amount of time the equipment is used. For example, a device that consumes 1,000 watts at any given time and runs for eight hours per day uses 8,000 watt-hours per day, or 40,000 watt-hours per five-day work week, or 176,000 watt-hours per month (a five-day work week will average 22 working days per month).

Finally, because you are probably billed in kilowatt-hours and not watt-hours, convert watt-hours to kilowatt-hours. To do this, divide your watt-hours by 1,000 to get kilowatt-hours. For example, if you are using 176,000 watt-hours per month, divide that by 1,000 and you get 176 kilowatt-hours (kWh). Once you have determined the number of kWh the device is consuming per month, note it on your worksheet in the "Quantity" column as shown in Figure 3.3.

You now know how many kWh the equipment consumes over a given period of time. Next, you need to determine how much that device costs you to run. In order to do this, determine the price that you are paying per kilowatt-hour. This information is usually provided on your hydro or electric bill. If you are having trouble finding it, contact your provider, ask what your kWh rate is and they can quickly tell you the rate.

Once you have determined your kWh rate, multiply the kWh rate by the amount of kWh you are consuming and you will know the cost of running that device for a given period of time. So, for example, if you determined that during a one-month period a device consumes 176 kWh and your kWh rate is 10 cents per kWh, then 176 × $0.1 is $17.60 per month to operate. Once you have determined the cost of running the device, note it on the worksheet in the column labeled "Cost" as shown in Figure 3.3. Note that $17.60 is not much money, but when you add up all the devices consuming energy, plus those devices that are much larger consumers of energy, such as large machines and ovens, you will quickly see the huge costs associated with energy consumption and where those costs lie in your value stream.

This method of measuring your energy consumption works well for getting a grasp of the amount of energy a device consumes and the associated costs of consuming that energy relative to your other devices. This method is widely accepted as a means of calculating the energy use and cost of a device. Using this method allows you to focus on the largest offenders (the energy-sucking equipment) when you proceed to the next step of minimizing energy usage to cut your costs and environmental impact. It is perfectly acceptable to use this method. Using the nameplate method, however, does not give you an exact, precise measurement. The reason for this is that the nameplate rating listed on the device is the maximum amount of power that device will consume at any given time, and some items may not consume that maximum amount all the time. If your 100-watt device consumes 100 watts half of the time and 90 watts the rest of the time, this difference is not taken into consideration with the nameplate method. The only way to get an exact, precise measurement of how much energy a device is consuming is to measure it with an energy meter.

Using the Energy Meter Method

There are many energy meters on the market, including ones that range from under $50, like the "Kill A Watt" meter for measuring simple electrical devices (like computers), to ones that cost thousands of dollars. The more expensive meters are more accurate in their readings and can read multiple devices at once, plus they have the ability to communicate the data to different software programs. However, they also require more technical knowledge and effort to operate. If you want an exact reading of energy consumption and you have neither the meter itself (although you can rent them) nor the experience required to operate one of these meters, have an in-house electrical engineer or consulting energy professional do this measurement for you.

Whatever method you use to measure energy consumption, ensure that you have done this for each item that you have identified for each activity of the value stream and of the overall building, and note the amounts on the worksheet in Figure 3.3. When you're finished, to check whether your measurements are accurate, add up the total costs of all activities and the overall building, and compare that cost to your monthly electric or gas bills.

CHOOSING AN ENERGY CONSULTANT

Before hiring an energy consultant, contact the local utility that sends you a monthly bill to see whether it can do energy consumption measurements for you. Many energy companies provide an "energy audit" for free. If that is not possible, look for an energy auditor or manager who is qualified and certified, such as a certified energy manager (CEM) under the Association of Energy Engineers (AEE) certification program. Your local chapter of the AEE will be able to assist you in finding the right person to do this measurement for you. You can also do an Internet search for "energy manager" or "energy measurement services in [your area]."

Another way to determine a precise measurement for a particular piece of equipment is to contact a supplier of a more energy-efficient model of the device you want to measure, and that company may be able to come in and measure it for you at little or no cost, in hopes that you will replace your existing device with one of their more energy-efficient devices.

If the two figures are close, you know that you have correctly measured, but if they are way off, you have either missed something or made an error in your calculations, and you need to go back and have another look.

The Greanco green team had already identified the equipment that was consuming energy, but team members still needed to measure the amount of energy each item was consuming. To do this, they started by determining how much energy each item consumed at any given moment. For most items, they were able to read the nameplate that listed either the watts, kilowatts, amps, volts, BTUs, or horsepower. For some items, they could not find or read the nameplate, so they had to call the manufacturer for the information. They then had to figure out the length of time each item was in use. They decided to get the usage time per day, and then multiply by 22 (the average number of working days, Monday through Friday, per month) to get a monthly average. For most items, they were able to ask the workers how long per day the item was used or turned on. They had to time the usage of other items themselves and determine an average, but, for the most part, finding out the average time used per day was fairly easy.

Next, they determined monthly consumption in terms of kWh, following the nameplate method to do this for each item. After the first couple of times, they were able to do this calculation quickly and easily. Finally, they needed to figure out the cost of consumption. To find their kWh rate, they called the utility that sent them a monthly bill to ask, and they found out that it was 10.5 cents per kWh. They noted all this information on their worksheet, as shown in Figure 3.3. A similar approach was followed for the overall building, and you can see the results of this in their building worksheet shown in Figure 3.4.

When it came to items that were common in the overall building such as IT equipment, lighting, and HVAC, the team decided to add up all individual items and list the group total on their worksheet. If they needed to, they could go back to their notes to see what made up that total number.

Energy Waste Elimination Worksheet					
Activity or Area:		***Overall Building***			
Current State					
Identify		**Measure**			
				Consumption	
Item	Source	Rate	Usage	Quantity	Cost
IT equipment	*Electricity**	*4200W*	*24 hrs/day*	*2217 kW hrs/mth*	*$232.79*
Lighting	*Electricity**	*6600W*	*18 hrs/day*	*2288 kW hrs/mth*	*$240.24*
Test equipment	*Electricity**	*1200W*	*1 hr/day*	*26 kW hrs/mth*	*$ 2.73*
HVAC	*Electricity*/gas*	*3000W / 50M3*	*20 hrs/day*	*1320 kW hrs/mth*	*$138.60*
* Electricity provided from a coal-fired generating plant					

Figure 3.4 Energy waste elimination worksheet. Current state: overall building.

Step 3: Minimize the Use of Energy

Now you are ready to start saving some money and the environment. Given that you now know each of the items that are consuming energy in the value stream and in the overall building, focus on those items that are consuming the most energy. Start to minimize the amount of energy they consume, so that you realize both the economic and environmental savings that come from minimizing energy use. The economic savings are simple and stem from the fact that you pay according to how much energy you use, so using less energy means you pay less. The environmental benefits stem from the fact that the energy you use needs to be generated say by coal or gas, and these pollute the environment by creating and discharging greenhouse gases (GHG) such as carbon dioxide (CO_2) into the atmosphere. Using less energy means that less GHG is discharged into the atmosphere. If you are using energy from a source with no emissions, than the reduction of energy means there is more of this cleaner energy to go around which is providing a positive impact on the environment. There are three broad things that you can do to minimize energy waste: practice energy conservation, use energy-efficient technologies, and practice energy management. These three techniques can be used for value stream activities as well as the overall building. We will look at each technique in turn.

Energy Conservation

Energy conservation in this context refers to using less energy by physically using an energy-consuming device for less time, or by not making it work as hard. Some classic examples of this type of minimization include turning off the lights when you leave a room, closing a window in the winter because "you're not heating the outside" (as your mother used to tell you), not standing with the refrigerator door open, or turning off your computer when you go home for the day. These are all examples of energy conservation, and you can use this technique to help you minimize the amount of energy being wasted. This is done by using energy-consuming devices only when needed, at the level that is needed.

Given that you already know your usage of each energy-consuming device, as determined in the Measure step, next you need to determine how much you actually need to use that device. For example, if you identified a machine that consumes energy and that machine is consuming energy for, say, eight hours a day, you should ask whether it actually needs to be on or consuming energy for eight hours a day. Devices that are on and consuming energy for a whole day may be in active use for only one, two, or three hours per day; the rest of the time the device is waiting to do some work. To determine the required usage of each device, ask the people who use that device to track how much they actually need to use it as opposed to how long it is on. Once you have determined the required usage time, you can move toward having that device consume energy only for the required usage time (one, two, or three hours, as opposed to eight). There are a few ways to approach this.

First, you can do it manually by shutting off the machine when it is not being used. Developing a policy and procedure for this and training the workers who use that machine to follow that policy and procedure is a good way to ensure that the machine will only be consuming energy when needed. Using signage can also help, such as posting a "Turn off machine when not in use" or "Turn off lights when leaving the room" sign. You can also use technology to automate or help you in conserving energy. There are many products on the market that can sense when an energy-consuming device is not being used and automatically shut it off. Such products include motion-sensing lights, occupancy sensors, devices that automatically shut off after prescribed periods of inactivity, and simple products such as timers. The point is to make sure that each energy-consuming device is on and consuming power only when required and is turned off at other times.

To calculate your savings from this form of conservation, recalculate your measurements with the new required usage time to determine the improved energy consumption from that device, and then subtract that result from your old energy consumption. For example, a device that pulls 100 W at any given time and runs for eight hours a day uses 800 watt-hours per day (or 24,000 watt-hours or 24 kWh per month). If you were able to conserve energy by running that device for only three hours a day, then 100 W × three hours a day results in 300 watt-hours per day × 30 days in a month, which is 9,000 watt-hours or 9 kWh per month. The difference, 24 kWh – 9 kWh, is a savings of 15 kWh. Finally, 15 kWh × 10 cents per kWh is $1.50 in savings per month, for little effort.

Be careful, because even when a device is turned off, it does not mean that it is no longer consuming any energy. When an energy-consuming device is turned off, it may still be consuming a little bit of power, known as "phantom power." A simple example is that when your TV is turned off, it is not actually off but rather on standby, waiting for a signal from the remote to turn back on. Although it is consuming a much smaller amount of power compared to when it is on and running, it is still consuming some power. To find out whether a device is consuming energy when turned off or has phantom power, use an energy meter or energy auditor, which will tell you the amount of power that is being consumed when the device is turned off. Another way to find out whether an item has phantom power is to do an Internet search on "phantom power of [device in question]." You can do a couple of things to eliminate this phantom power. One is to unplug the device from the wall; although this works, it can be a hassle. Another way is to use power bars or power strips and turn the power bar off when the device is not in use. This will cut all power going to the device and eliminate the phantom power load, saving electricity and the cost of that electricity. There are also smart power bars or strips that can be programmed to turn on and off or have motion sensors that turn the strip on or off.

Another way to conserve energy is to lower the intensity of the device by, for example, lowering the temperature. Does the heat or the oven need to be that hot? Could you turn it down by one or more degrees? If so, you will save a considerable amount of energy. Or can you turn the temperature of a cooling device such as an air conditioner, refrigerator, or freezer up one or more degrees? Again, this will save a considerable amount of energy and money. Does a motor have to be run at full speed, or could you slow it down? We all know driving slower saves gas; the same principle works with other energy-consuming devices that have the ability to change speeds. You can achieve this by manually lowering or raising temperature or speed settings, or you can use technology to do it for you, for instance, by using heat and temperature sensors to control temperature so that the equipment is not running full blast, all the time. Programmable thermostats for heating and cooling work very well for the overall building.

Once you have developed a solution for ensuring that the identified energy-consuming devices are on and consuming energy only when needed and at the level that is needed, and after you have addressed phantom power issues, note the solutions in the future-state section of your Energy Waste Elimination Worksheet, as shown in Figure 3.5 and Figure 3.6, including the savings that go along with the solution.

Energy Waste Elimination Worksheet						
Activity or Area:			***Value Stream Activity 1: Forming***			
Future State						
Minimization						
Item	Conservation		Efficiency		Management	
	Solution	Savings	Solution	Savings	Solution	Savings
Forming machine	*Turn off when not in use*	*$235.62/ mth*				
Crane						
Computers						
Workstation lights						

Figure 3.5 Energy waste elimination worksheet. Future state: value stream activity 1.

Energy Waste Elimination Worksheet						
Activity or Area:			***Overall Building***			
Future State						
Minimization						
Item	Conservation		Efficiency		Management	
	Solution	Savings	Solution	Savings	Solution	Savings
IT equipment	*Put on timers*	*$58.21/ mth*				
Lighting						
Test equipment						
HVAC						

Figure 3.6 Energy waste elimination worksheet. Future state: overall building.

The Greanco green team was now ready to look for ways to reduce their consumption of energy so that they could save some money and the environment. What first jumped out at them was the forming machine. Looking specifically at how they could use the technique of conservation, the team focused on this machine. They knew that currently it was running for the total length of the two shifts, and was turned off at night. The first question the team had was, "Is the machine actually being used for the full 16 hours per day?" The expected answer came back: "No, it is only used for about 20 minutes every hour; the rest of the time it is being loaded or cleaned, or it's break or lunch time." The team then asked if it would be inconvenient to turn it on and off regularly. The workers replied that it takes about 15 to 20 minutes for the heater to come up to temperature. The team discovered that, because the heater was fed by gas, they could, with a few minor adjustments, shut off the machine but leave the heater on. They were able to cut the usage by two-thirds: Instead of running 16 hours per day, the forming machine could run for just 6 hours a day. The savings were significant, as noted on their worksheet in Figure 3.5.

When it came to the overall building, the Greanco green team decided to focus on the IT equipment. The obvious question was, "Why is IT equipment, such as computers and faxes, running 24 hours per day if the company only runs two shifts, or 16 hours per day?" The rather frustrating answer was that people didn't make the effort to shut down their computers or other IT equipment before leaving. The Greanco green team knew that if they could convince people to shut off their IT equipment every day, they could save a great deal of energy and money. With permission from senior management, they rolled out a new policy requiring people to shut off their equipment before leaving. This program worked modestly and achieved some savings, but people were still forgetting or not bothering to do it. The Greanco green team decided that, instead, they would put all IT equipment on simple timers that would shut it off one hour after the last shift and turn it back on one hour before the first shift. This saved six hours of usage every day. The team noted the significant savings on their worksheet in the "Conservation" column, as shown in Figure 3.6.

Energy-Efficient Technologies

The next way to minimize energy waste is to use energy-efficient products to reduce the amount of energy consumed when the device is turned on. An energy-efficient product is one that provides the same service or function as a traditional product but uses less energy to do so. A well-known example is energy-efficient lights. A typical incandescent lightbulb uses 100 W to produce a certain amount of light (known as lumen output), whereas an energy-efficient lightbulb uses only 23 W to produce the same lumen output. The savings is 100 W – 23 W, or 77 W. To calculate the savings from using an energy-efficient product, recalculate your energy measurements using the new rate of energy being consumed at any given time. In our example, a 100-W device used for eight hours per day consumes 800 watt-hours per day, or 24,000 watt-hours (24 kWh) per month. The new energy-efficient product that uses only 23 W, even if used for eight hours, consumes only 184 watt-hours per day (23 W × 8 hours), or 5,520 watt-hours (5.52 kWh) per month. The difference in consumption, 24 kWh – 5.52 kWh, is a savings of 18.48 kWh. At 10 cents per kWh, the cost savings per month is $1.85.

Some energy-efficient products do cost more up front, but the savings usually pay back that up-front investment very quickly, providing ongoing savings for the life of the product. Considering the cost of operating a product over its life span is known as life-cycle costing. Using life-cycle costing, as opposed to the simple up-front cost, to determine which product you should purchase is always recommended. One product may be cheaper to purchase up-front compared to a more energy-efficient product, but the cost of operating it usually far outweighs anything you save on the up-front cost.

Over the past decade, the emergence and evolution of energy-efficient products has been astonishing. There are energy-efficient products for almost any device you can think of, from lights to computers to motors. Oftentimes, there are different levels of energy efficiency for each product as well, so that if you are constrained by budgets, you can choose the most energy-efficient product that fits your budget. Suppliers of a particular item will often be able to show you a more energy-efficient version. If that supplier does not offer a more efficient model, do an Internet search on "energy efficient [device in question]," or contact your local utility or government for help. Many suppliers now offer free services, whereby they come to your location and educate you on the latest energy-efficient products available. Once you have identified the energy-efficient solution that works for each item that is consuming energy, note it on the worksheet in the "Efficiency" column, as shown in Figure 3.7, along with the savings you will get.

When it came to energy-efficient products, the Greanco team wanted to focus on the lowest-hanging fruit to prove to senior management that this works. They looked at lighting first, and instead of focusing on the value stream and overall building separately, they decided to address all lighting at once. To explore the possibilities, the team called in a lighting retrofit vendor who looked at the lighting in both the value stream and the overall building. The vendor was able to assess the situation quickly and provide the cost of retrofit and the payback period. In short, the team discovered that it could save about 25 percent of the lighting costs with a simple payback of approximately two years. With rebates that were available, the payback dropped to 1.5 years, with immediate positive cash flow if they were to use a credit line with minimum monthly payments to finance the project. This was a no-brainer. Management immediately approved the expenditure, and within one month's time they were realizing the positive cash flow. Within a year and a half, they would achieve the monthly savings shown in the "Efficiency" column on the worksheet in Figure 3.7 for the value stream activities and the worksheet in Figure 3.8 for the overall building.

CASE STUDY: THE WEYERHAEUSER COMPANY

The Weyerhaeuser Company, one of America's leading forest products producers, gives top priority to environmental issues. This regularly wins the company awards from government agencies, industry groups, the financial community, and environmental organizations. In an effort to reduce energy consumption from the use of motors in its operations, Weyerhaeuser put in place a corporatewide policy for the replacement of motors with more energy-efficient ones. The policy was based on a detailed study that was done to determine the best way to reduce the total cost of ownership for the motors. Not surprisingly, the option that had the greatest economic benefits also had the greatest environmental benefits. By using motors that are more efficient than regulated by legislation, the company not only helps save the environment by using less energy—thus preventing the environmental impacts associated with energy use—but also saves the money it would have had to pay for that energy. Weyerhaeuser changes out enough motors every year to save almost $400,000 a year in energy costs. The company can keep changing out motors at this rate for a number of years until all its motors are replaced. This means that the company can increase its savings every year by $400,000 until all its motors have been replaced. For example, in year two the company will save $800,000 to add to the $400,000 from year one that it has already pocketed. Now that is savings!

Energy Waste Elimination Worksheet						
Activity or Area:			***Overall Building***			
Future State						
Minimization						
Item	Conservation		Efficiency		Management	
	Solution	Savings	Solution	Savings	Solution	Savings
Forming machine	*Turn off when not in use*	*$235.62/ mth*				
Crane						
Computers						
Workstation lights			*Replace with T8 bulbs*	*$14.78/ mth*		

Figure 3.7 Energy waste elimination worksheet. Future state: value stream activity 1.

Energy Waste Elimination Worksheet						
Activity or Area:			***Overall Building***			
Future State						
Minimization						
Item	Conservation		Efficiency		Management	
	Solution	Savings	Solution	Savings	Solution	Savings
IT equipment	*Put on timers*	*$58.21/ mth*				
Lighting			*Replace with T8 bulbs*	*$60.06/ mth*		
Test equipment						
HVAC						

Figure 3.8 Energy waste elimination worksheet. Future state: overall building.

It is important to note that there are a number of creative ways to finance the purchase of an energy-efficient product that will replace or retrofit an existing product. One way is a simple lease or “mortgage” of the item. In many instances, the savings from an energy-efficient product will be larger than the lease or mortgage payments. This provides a positive cash flow immediately, with even more cash flow improvement and savings once the item has reached its payback period. Another way to finance is to use "performance contractors" or energy service companies (ESCOs). These contractors will purchase the item and install it. In turn, you pay them with your savings in energy charges until the item is paid off, at which time the item is yours. You realize the savings without ever having to make an up-front investment or negatively affect cash flow.

When planning new purchases that will not replace an existing device, these options are not possible. You will need to do the life-cycle costing assessment and make the most intelligent choice within your limits. However, for the purchase of both new energy-efficient products as well as energy-efficient products to replace an existing item, there are a number of incentives and rebates available to help offset the cost of purchasing more energy-efficient products. These rebates and incentives typically come from the utility company or government and range from hard dollars back per device, to a percentage-of-cost rebate, to tax breaks and tax write-offs. To find these incentives and rebates, contact the local utility that sends you your monthly bill and/or your government. Most cities in North America have some rebate or incentive for purchasing energy-efficient products. You can also do an Internet search on "rebates in [your area] for energy efficient [device in question]." There are also rebates and incentives for practicing simple conservation and management techniques, which we will discuss in the following section. The utility or government will assess your current and previous energy usage and patterns, and provide rebates or incentives for improvement.

Energy Management

The final way to minimize energy waste is through energy management. This technique does not actually reduce your energy consumption per se, but rather your energy cost. Remember that energy waste is not only the use of energy, but also the associated costs. To use this minimization technique, you will first need to understand how you are billed.

Typically, your energy bill consists of two basic charges: The first charge is a consumption charge, a rate per kWh. For example, if you use 1,000 kWh per month and your kWh rate is 10 cents, your one charge is 1,000 kWh × 10 cents, equaling $100. In many cases, this charge may fluctuate based on the time of day you are consuming energy, and may be higher during peak periods. If your charge is usually 10 cents per kWh, during peak times it may jump to 12 or 15 cents. In addition to this consumption charge, you are usually charged a second fee, known as a demand charge or demand penalty. This charge is based on your largest one-time consumption of energy in a given time period, usually a 15-minute window. For example, if you used 1,000 kWh in a month and at one point in that month you pulled 10 kW all at once, even for a couple of minutes, because all your machines were running at the same time, you are charged a demand penalty or fee on that 10-kW one-time demand. This fee is usually higher than your consumption charge, and can be $3, $4, or $5 (or higher) per kW. Your total charge is then $100 for consuming 1,000 kWh in the month at 10 cents per kWh, plus the demand charge of $40 (10 kW × $4), for a total bill of $140. (If you have trouble dissecting your energy charges, contact the utility that sends you your bill; they can help you understand your unique situation better.)

You can see that by lowering your one-time peak demand (the 10 kW), you would save on the demand charge. For example, if you lowered your peak demand to 8 kW, your demand charge would be reduced to $32 (8 kW × $4). By managing your peak demands and the times of day that you consume energy, you can lower your energy costs even if you are still consuming the same amount of energy.

You can approach energy management in a couple of different ways. First, if you are charged higher rates based on the time of day, manage this by moving energy use to a time of day when the charge is lower. Doing this may involve changing work times or schedules, or putting devices on timers to turn on at night (for example, an oven) so that you consume energy during the time of day that has the lowest charges. Second, lower the peak demand of energy use by managing the amount of energy that is used at any one time. Doing this may, again, involve changing work schedules and times or putting devices on timers to turn on at night. It may involve staggering production activities. For example, running ten machines in your value stream all at the same time means demanding a lot of energy at once, as opposed to running one or two machines (or anything less than ten machines) at one time.

Staggering production activities can get rather complicated as you try to work around production schedules, customer requirements, and so on, but it can be done, and the rewards are worth the effort. Wherever possible, defer running machines or consuming energy to times when the least amount of energy is being used, generally at night or early morning, depending on your shifts. If you operate 24 hours a day, you will need to figure out how you can stagger the use of energy so that consumption is spread out instead

Energy Waste Elimination Worksheet						
Activity or Area:			***Overall Building***			
Future State						
Minimization						
Item	Conservation		Efficiency		Management	
	Solution	Savings	Solution	Savings	Solution	Savings
Forming machine	*Turn off when not in use*	*$235.62/ mth*			*Load sheets before turning on machine*	*$25.20/ mth*
Crane						
Computers						
Workstation lights			*Replace with T8 bulbs*	*$14.78/ mth*		

Figure 3.9 Energy waste elimination worksheet. Future state: value stream activity 1.

of being consumed all at once. For help in doing this, approach your utility or government; in many cases, they will do this for free. You can also approach a third-party energy manager (such as a certified energy manager [CEM] through the Association of Energy Engineers [AEE]) or another energy management company that has experience in doing this. To find them, do an Internet search for "energy management companies in [your area]." Once you have developed a way to reduce peak demand and/or use energy during the cheapest time of day, note it on your worksheet in the "Management" column, as shown in Figure 3.9, along with your savings.

Of course, using the first two techniques of conservation and energy efficiency will contribute to lowering peak demand and provide savings on both the consumption charge and the demand charge.

The Greanco green team was now looking for ways to cut energy costs through strategic management of their energy consumption patterns. Members set out to understand how they were billed. Trying to dissect the bill themselves was producing no results, so they called the utility and asked them to explain how they were being charged. They found out that they were paying a demand charge of $4.20 per kW for their largest one-time consumption of energy at any given point, but not being charged higher rates at certain times of the day because their consumption was not large enough for them to be penalized in that way. Therefore, the team focused on how they could reduce their peak demand.

Looking at only the first activity in their value stream, they could see that the crane and the forming machine were the largest consumers of energy. If both were on at the same time, a lot of energy would be pulled at once in order to operate both devices. In examining the process, they noticed workers turning off the machine when not in use, which was good, but then turning it on and loading it using the crane. Both pieces of equipment were consuming power at the same time, causing a major spike in energy consumption. The team asked whether the machine could be loaded first, and then turned on. The workers didn't see any problem with that, and this was confirmed after an inspection by the operations engineer. They put a new procedure in place right away and shaved their peak demand by 6 kW.

Previously, their peak demand included the normal daytime running consumption (lights, HVAC, IT, forming, crane, etc.). Once they stopped adding the crane to the peak consumption, the peak demand dropped by the amount of energy consumed by the crane, 6 kW. Because their peak demand charge was $4.20, the savings was $25.20 per month or $302.40 per year. Not bad for a quick, simple, no-cost change. See Figure 3.9.

Step 4: Offset Remaining Energy Use

Offsetting is a technique that is used by many organizations to counteract or neutralize the environmental impact caused by the greenhouse gas emissions that result from their use of energy. A majority of the energy we consume comes from generating plants that use nonrenewable natural resources to create energy, such as burning coal. These nonrenewable sources of energy release large amounts of greenhouse gas emissions into the atmosphere that contribute to pollution, smog, and the overall problem of global warming.

One way to neutralize or counteract these emissions is by contributing to projects that will remove or prevent emissions from being released into the atmosphere. Typically, these projects involve building renewable sources of energy such as wind and solar farms, or reforestation projects. These projects are funded solely through the sale of offset tickets or certificates.

Because these projects would not occur without the sale of these offset certificates, it is fair to say that, by purchasing these offsets, you are contributing to removing or preventing greenhouse gas emissions from being discharged into the atmosphere. By subtracting the emissions that are prevented or removed from the atmosphere through the purchase of offsets from the emissions caused by your energy consumption, you are in theory reducing your environmental impact. In other words, you are offsetting your environmental impact.

Offsetting emissions caused by your energy consumption results in a number of benefits beyond environmental ones. Although it does cost money to purchase these offsets, the indirect benefits are fairly substantial. If you promote the fact that you offset your emissions, it will show customers and employees that you are indeed committed to the environment, resulting in increased employee/customer attraction and retention. Many consumers and employees actively seek organizations that are doing this, and use the information to help them choose who to support or join. If you are subject to some sort of carbon tax, it may also be possible to use the purchase of carbon offsets to bring you below the threshold where taxes kick in, thus saving you those taxes or penalties and fees. The cost of purchasing these offsets is often eligible as a tax write-off, and there are other forms of incentives that can offset the cost of your offsets. Of course, offsetting should be seen as a temporary solution after minimization, to bridge the gap to using self-harvested renewable energy. It is a solution that does not address the root cause: the generation of emissions in the first place from the use of fossil fuels.

In order to offset your emissions from the use of energy, you first need to know the quantity of emissions being generated as a result of your energy consumption. To find this out, you need to know how much energy you are consuming and the source of that energy. Because this has already been done in the Measure step, it is easy to determine the amount of emissions generated as a result of your energy consumption. There are numerous carbon-offset vendors that have free calculators on their Web sites that will calculate the amount of GHG emissions (usually expressed in tons of CO_2 or CO_2 equivalents [CO_2e]) that are discharged into the atmosphere as a result of your energy consumption. All you need to do is plug in the amount of energy you are using (for electricity the kWh; for fuel sources like natural gas, the amount you are using in meters cubed [m^3]) and click the calculate button. The calculator figures the amount of emissions and tells you how

Energy Waste Elimination Worksheet				
Activity or Area:		***Overall Building***		
Green State				
Elimination				
Item	Offsetting	Renewable Energy		
		Solution	Cost	Savings
IT equipment	*$90/mth*			
Lighting				
Test equipment				
HVAC				

Figure 3.10 Energy waste elimination worksheet. Green state: overall building

much it would cost to purchase enough offsets to completely neutralize or "offset" the emissions you have created through your use of energy.

If you have a special circumstance, for instance, if you are using a specialized fuel, contact the vendor, explain your situation, and ask them to calculate the amount of emissions generated and the number of offsets needed to neutralize those emissions. In order to fully neutralize or offset your emissions, you must purchase the full amount of offsets; anything less results in less than all your emissions being offset, leaving you short of neutrality. If constrained by budgets, you may choose to offset only your value stream activities or only the overall building. But it is highly recommended that if you are going to offset, you offset as much as possible in order to be as close as possible to neutrality, because this will bring you the greatest rewards in terms of attracting new customers and employees. To find these offset vendors, do an Internet search for "carbon offsets" or "carbon offset vendors," and numerous results will pop up. Once you have determined the amount of emissions being generated, the number of offsets needed, and the cost of these offsets by using the free calculators, note it in the Green State section of your Energy Waste Elimination Worksheet, as shown in Figure 3.10.

Having focused primarily on the value stream activities, the Greanco green team decided to explore the cost of offsetting the overall building's environmental impact from the consumption of energy. They looked up a couple of carbon offsetting vendors and found one that was offering Gold Standard Offsets and had a user-friendly calculator available. They added up the total monthly consumption of energy, which was 5,851 kWh, plugged that number into the calculator, and discovered that it would cost them approximately $90 per month to offset this amount. They noted it in their worksheet, as shown in Figure 3.10, and went to work on seeing if they could use some of the savings from the minimization efforts to purchase the offsets.

When choosing an offset vendor, there are a number of things to look for: verification that the projects are real and actually taking place, verification that these projects would not have happened without funding from the sales of the offset certificates, information on whether the offset project reduces our dependency on fossil fuels, and so on. An organization called Clean Air Cool Planet has published a guide entitled "Consumer's Guide to Carbon Offsets for Carbon Neutrality" that will aid you in determining the offset vendor that is right for you. There are also a number of standards for carbon offsets, such as Green-e and the Gold Standard. The Gold Standard is probably the highest and most recognized standard for carbon offsets and is supported by a number of nongovernmental organizations (NGOs) such as World Wildlife Fund, Greenpeace, and the David Suzuki Foundation. It is recommended that you work with a vendor that provides offsets that are verified by the Gold Standard or another highly recognized standard to ensure that the offsets are having the most positive impact on the environment.

Step 5: Transition to Self-Harvested Renewable Energy

The final step—and the end goal of eliminating energy waste—is to move toward self-harvested renewable energy. Renewable energy means sources of energy that are rapidly replaced once used and that result in little or no emissions being discharged into the atmosphere. This includes sources such as:

- Solar power, coming from the use of solar panels or photovoltaic panels, which convert heat from the sun to generate usable energy

- Wind power, which generates usable energy from the turning of blades on a windmill
- Geothermal, which uses the heat that lies below the Earth's crust to generate energy
- Biomass, which burns biological materials (with no emissions) that are rapidly renewed to create energy
- Hydro or microhydro, both of which use the flow of water to spin a turbine and create energy

Having your energy come from renewable energy sources is a great thing, but the self-harvesting part of it is where you will get the most benefits. If you are purchasing green power from renewable energy providers, the fact is, you are still paying for your energy, which is part of the energy waste. You could, on the other hand, be getting it for free from the sun, wind, or other renewable sources by being able to harvest it yourself. In this final step, you will aim to invest in the ability to harvest your own power for free from the sun, wind, earth, or water. When you can satisfy 100 percent of your energy needs through self-harvested renewable energy, you have completed this last step and eliminated energy waste completely.

Now imagine the end result: You no longer have any environmental impact as a result of your energy consumption, and you no longer have to pay for energy. This is huge. With the current high cost of energy and guaranteed increases for the future, imagine the direct cost savings. It is tens or hundreds of thousands of dollars annually for small and medium-sized enterprises, and even millions of dollars for larger enterprises. In addition, imagine the great press that would come from this, resulting in the attraction of new customers and employees, the loyalty of existing ones, and the increased perception of value from shareholders. Adding all these benefits together provides ample justification for this last step that cannot be ignored if you truly want to be green and get the most economic and environmental benefits.

The first step in being able to self-harvest your own renewable energy from the sun, wind, or other sources has already been addressed: It is to minimize the amount of energy that is being consumed. You could go straight to this last step instead, if you wanted, but then you would be increasing the size and cost of the system that will need to be installed, as opposed to minimizing the amount of energy you consume first so that you can minimize the size and cost of the system that will need to be installed.

The second part of the process is to decide on what form of renewable energy you could self-harvest at your location. If you are in a location that

CASE STUDY: GOOGLE

Google is one of the most recognized, forward-thinking, and fastest-growing companies around today. Being the Internet's premiere search engine, they require an enormous amount of energy to operate the computers and servers they use to run their business. As an emerging leader in environmental stewardship, they recognized the need and opportunity to eliminate energy as a waste and move toward self-generated clean energy. Their focus was on their Mountain View, California, location known as the Googleplex, which because of its unique architecture, posed a number of challenges. Overcoming those challenges, they covered 197,000 square feet of their corporate campus with solar photovoltaic (PV) cells. By building the largest solar power system ever installed at a single corporate campus, they will save more than $393,000 in annual energy costs with a payback of 7.5 years. Any increase in energy costs, which is inevitable, will increase savings and reduce payback time. Google plans to increase its amount of self-generated clean energy as well as tackle its other green wastes.

receives a lot of sun, then perhaps installing solar panels on your roof or on stands covering parking areas will be the best solution. If you are in an area that is consistently windy, the erection of a windmill on-site might be the best solution. If you are located close to a moving body of water such as a river or stream, then maybe the installation of hydro turbines will be the best solution.

The way to determine the best solution for you is to have a feasibility study done by suppliers of renewable energy systems. These organizations will be able to quantify the amount of power you could generate from wind, solar, and geothermal and inform you of the up-front costs as well as the payback periods involved. It is important in exploring each option that you look into the financing options available. Payback periods may be lengthy, but leasing or mortgaging the system may provide immediate positive cash flow with little or no money up front and out of pocket. With information on the amount of energy that can be generated at your site, the cost of installation, payback periods, and cash flow scenarios, you can then make a decision on the best solution for you.

Chapter 4

The Second Green Waste: Water

Just like energy, water is an integral part of operating any business. For some, water may play a much more important role in operations, but it is definitely a requirement for any business. Whether we use water in the production process, or use it in the running of an office or building, we all use it and need it. There are attractive, if not necessarily well-known, economic benefits to tackling this waste.

First, we need to understand that the waste of water is not necessarily in using it, but rather having to pay continually rising prices to consume it and to discharge contaminated water, both of which have a negative environmental impact. It is easy to see the economic benefits of eliminating this waste—the savings realized from not having to pay for the water that is consumed and the reduced levies and fines associated with discharging contaminated water. But often overlooked is the benefit of addressing the fact that there's only so much water, and it is becomingly increasingly scarce, which continually drives up the price. If we can't get water, we will have extreme difficulties running any aspect of a business or, at the very least, be subject to continually increased costs. You can count on the fact that in the not-too-distant future, water will become as big a crisis, if not bigger, than the current oil crisis. This is why the ultimate goal should be to get as much of it for free from rainwater harvesting, and then to reuse as much as possible.

If you're getting water for free from rain and reusing what you can, you are inherently minimizing or eliminating the costs associated with paying someone else to deliver it to you and take it away to be cleaned. Imagine the end result: You draw the minimum amount of water for free, and then use it over and over again for free. Of course, it can take time to achieve this, which is why we will first look at ways to produce immediate benefits by minimizing water consumption.

As with all waste elimination processes, you will use a worksheet to help work through the steps:

Step 1: Identify the source and use of water in your value stream and in the overall building
Step 2: Measure the amount of water consumed in each activity and in the overall building
Step 3: Measure the toxicity of water discharged in value stream activities
Step 4: Minimize the amount of water used both in the overall building and within your value stream
Step 5: Minimize the toxicity of the water discharged
Step 6: Self-harvest rainwater
Step 7: Transition toward continual reuse of water

INTENDING TO REUSE WATER

If you intend to reuse water, all the activities you do leading up to that will align with and cater to the end goal of water reuse. This approach will give meaning to all the little day-to-day steps you need to take in order to get to reuse. With the intention of achieving this end goal in mind, you are ready to get started.

Step 1: Identify the Use of Water in the Value Stream and Overall Building

In this step, you should not worry about the amount of water used; we will capture that in the next step. Rather, be concerned about identifying the use of water within your value stream and in the overall building. Start by looking at the activities of the value stream. You may already know some of the activities that are consuming water, but to ensure you identify all the activities that are consuming water, start at the first activity of the value stream. Walk to the area where this activity is being performed and look for anything that could potentially be dispensing water, such as taps, hoses, and sprayers. If you spot any of these, dig a little deeper by asking the employees who work at this activity whether water is being dispensed from these things. Remember that at this point you are not concerned with *how much* water is being used but that it *is* being used in that activity.

If none of these things are visible, this doesn't necessarily mean that there is no water being consumed in this activity, as it may be hidden. To make sure you don't miss anything, look and see whether there is any liquid in the process. If there is, what kind of liquid is it? You can find the answer to this question by asking (a) the employees working in the area or (b) the materials or purchasing managers. If it is just water, you have identified that there is indeed water being used in the process, and that is all you are aiming to do at this point. If it is not water, did the liquid come in as a liquid or was water added to a dry substance to make this liquid? If it came in as a liquid other than water, do not identify it as water waste; this liquid should be identified as a material input when looking at material waste. If, on the other hand, this liquid first arrived as a dry substance, you must have added water to it to make it a liquid, and you should identify it as the use of water.

If you don't see any signs of water being used at all, take one extra step and ask the employees working in the area or the production or plant manager if they know of any water being used in this activity (you can do this first, as it may avoid extra work). If they all answer "no," there is a pretty good chance that no water is being used. However, if you have identified water being used in this activity, note it by filling in the Current State section of the Water Waste Elimination Worksheet, as shown in Figure 4.1. You will need to repeat this for each value stream activity to identify the consumption of water in those activities. This should be a fairly simple and quick task; it should take no more than a few minutes to identify whether water is used at each activity.

You saw in Chapter 2 that the first activity (besides receiving) in Greanco's value stream is forming. In the forming process, the workers quickly spray the sheets of metal to get rid of any debris, and the forming machine itself injects water to keep the pieces of metal from overheating. Also located in the area is a washbasin that is used for washing dirty molds and tools. Greanco identified that these activities are consuming water and put that information into its worksheet, as shown in Figure 4.1.

Water Waste Elimination Worksheet					
Activity or Area:		***Value Stream Activity 1: Forming***			
Current State					
Identify	**Measure**				
Item	Flow Rate	Usage	Consumpton	Quantity	Cost
Sprayer 1					
Forming machine					
Wash basin					

Figure 4.1 Water waste elimination worksheet. Current state: value stream activity 1.

Next, we want to identify the use of water outside the value stream, in the overall building. This should be even easier than identifying water consumption in the value stream. Look for obvious items such as toilets, sinks, water coolers, and irrigation systems, and note the item and the quantity of each item. For example, if in your building or unit you have ten toilets, note that as shown on the worksheet in Figure 4.2. To make sure you capture things that aren't so obvious, speak with the facilities manager or the plant manager. You can also speak to the building owner or the landlord if you are having trouble identifying these items; they will be able to help you identify what is consuming water. Don't get too detailed at this point, though; remember that this should be a fairly easy and quick step. As you move through the steps and improve your skills, you will pick up anything you may have missed on the first pass.

Water Waste Elimination Worksheet					
Activity or Area:		***Overall Building***			
Current State					
Identify	**Measure**				
Item	Flow Rate	Usage	Consumpton	Discharge	Toxicity
Toilets (10)					
Sinks (8)					
Irrigation system					

Figure 4.2 Water waste elimination worksheet. Current state: overall building.

Greanco occupies a relatively small space of approximately 50,000 square feet. Throughout the building, the team was able to identify ten toilets, eight sinks, and an irrigation system that consume water. They filled in the worksheet, as shown in Figure 4.2.

Step 2: Measure Water Consumption and Discharge in the Overall Building and Value Stream Activities

Once you know the things that are consuming water, proceed to measuring the amount they are using. First, determine the flow of water coming out or, in other words, how much water is coming out per use or per period of time. Depending on the item, there are a number of ways to determine this. If a machine or piece of hardware is consuming water, it may have a nameplate or marking on it indicating how much water it consumes, like the stamp you see on all toilets that indicates how much water is used per flush. If it doesn't have a marking or you can't figure it out yourself, you can do a couple of things:

- Call the manufacturer of the item; they should be able to tell you.
- Crudely measure it yourself by getting a bucket of a specified size (one-gallon, five-gallon, etc.), and time how long it takes for the device to fill it. If it takes one minute to fill a five-gallon bucket, you know that it uses five gallons per minute.
- Ask your plumber, landlord, plant manager, facilities manager, or water provider for the information.
- Do an Internet search on "amount of water used by [name of device/process]." Don't worry about being exact down to the last drop; you're trying to get a good estimate of how much water each device is consuming so that you know what the biggest offenders are and can focus on minimizing those first.

Once you have determined the flow of water coming out per use or per period of time, note it beside the item, as shown in the worksheet in Figure 4.3.

Water Waste Elimination Worksheet					
Activity or Area:		***Value Stream Activity 1: Forming***			
Current State					
Identify	**Measure**				
Item	Flow Rate	Usage	Consumpton	Quantity	Cost
Sprayer 1	*3 gal/min*				
Forming machine	*0.25 gal/min*				
Wash basin	*1 gal/min*				

Figure 4.3 Water waste elimination worksheet. Current state: value stream activity 1.

The Greanco green team had no idea what the flow rates were for any of the items on their worksheet. For the sprayer, they timed how long it took to fill up a one-gallon bucket; it took 20 seconds. To determine the per-minute flow rate, they multiplied by 3 to get 3 gallons per minute, and filled in the worksheet, as shown in Figure 4.3. They performed the same exercise for the washbasin. When it came to the forming machine, they had no way of measuring it and could not find a marking on the machine that showed the flow rate. They called the manufacturer and discovered that the machine came from the factory with a set dispense rate of 0.25 gallons per minute, but that it could be adjusted to dispense more or less, depending on the thickness and type of metal being formed. They noted the flow rate in the worksheet, as shown in Figure 4.3.

When it came to the hardware in the building that was using water, it was quick and easy for the Greanco green team to determine flow rate. The toilets had the flow rate stamped right on them; being in an older building, they were using 6 gallons per flush. To determine the flow rate for the sinks, they performed the bucket exercise and came up with 1 gallon per minute. To determine the amount of water used by the irrigation system, they used the bucket trick on one sprinkler to determine the flow rate (it was 5 gallons per minute), then counted the number of sprinklers (there were 9) and multiplied that by 5 to get a rough estimate of 45 gallons per minute. They noted this information on their building worksheet, as shown in Figure 4.4.

Having determined the flow of an item, you will next be able to determine how much water it pulls in a day or month, or whatever timeframe you choose to measure, by determining the usage rate of that item. If you don't know the usage yourself, ask the employees that use the device or the facilities manager or plant manager how often it is used. For example, if there is a device that flows 3 gallons per minute and it is used for 60 minutes per day, it uses 180 gallons per day or 1,260 gallons per week. If it is a piece of hardware such as a toilet that uses so much water per use, you then need to determine the number of times it is used per day, week, or month. Once you know the usage, note it on the worksheet in the "Usage" column, as shown in Figure 4.5. Then you can multiply the flow rate times the usage to determine the consumption, and note that in the worksheet in the "Consumption" column. Just remember to keep the measurement timeframes consistent across all activities and devices. If you

Water Waste Elimination Worksheet					
Activity or Area:		***Overall Building***			
Current State					
Identify	**Measure**				
Item	Flow Rate	Usage	Consumpton	Discharge	Toxicity
Toilets (10)	*6 gal/flush*				
Sinks (8)	*1 gal/min*				
Irrigation system	*45 gal/min*				

Figure 4.4 **Water waste elimination worksheet. Current state: overall building.**

Water Waste Elimination Worksheet					
Activity or Area:		***Value Stream Activity 1: Forming***			
Current State					
Identify	**Measure**				
Item	Flow Rate	Usage	Consumpton	Quantity	Cost
Sprayer 1	*3 gal/min*	*60 mins/day*	*180 gal/day*		
Forming machine	*0.25 gal/min*	*420 mins/day*	*105 gal/day*		
Wash basin	*1 gal/min*	*90 mins/day*	*90 gal/day*		

Figure 4.5 **Water waste elimination worksheet. Current state: value stream activity 1.**

measure water consumption per day for one activity, all activities need to be measured per day.

If you are having trouble with this, bring in a water management specialist. They are generally available at minimal cost. To find such a specialist, call your local water conservation office or do a Google search on "water measurement or management specialists in [your area]." In most cases, however, you should be able to get a good idea by doing it yourself. A good check to make sure your measurements are accurate is to compare your measurements against your water bill. If your water bill says you are using 100,000 gallons per month and you only measured 50,000, you need to go and have another look.

At this point, you know where the most water is being used in your value stream, so you can focus your water-minimization efforts on activities that are consuming the most water.

The Greanco green team asked the employees who use the sprayer on a daily basis how often they use it and came up with an average of about an hour per day. They did the same for the forming machine and the washbasin, and then filled in the "Usage" and "Consumption" columns in the worksheet in Figure 4.5. In forming, the main consumer of water is the sprayer, so the team can focus on minimizing that.

When it came to the usage of the building hardware, they noticed that the sinks were used constantly during lunch hour and breaks, so they estimated that each sink tap ran for approximately two hours (or 120 minutes per day) and multiplied times eight sinks to get 960 minutes per day. At a flow rate of 1 gallon per minute, the total is 960 gallons per day for all eight sinks. Although they knew the water used per flush for toilets, they were unsure how many times they were flushed. They searched the Web for "average number of times a toilet is flushed per day" and came up with an average of about five times per day per person. With forty employees, that amounts to 200 flushes per day × 6 gallons per flush for a total of 1,200 gallons per day being flushed down the toilet (so to speak). The irrigation system was on a timer; it came on for one hour each morning and each evening for a total of two hours or 120 minutes per day × 45 gallons per minute, for a staggering 5,400 gallons per day. They noted all this information on their worksheet, as shown in Figure 4.6.

Water Waste Elimination Worksheet					
Activity or Area:		***Overall Building***			
Current State					
Identify	**Measure**				
Item	Flow Rate	Usage	Consumpton	Discharge	Toxicity
Toilets (10)	*6 gal/flush*	*200 flushes/ day*	*1200 gal/day*		
Sinks (8)	*1 gal/min*	*960 min/day*	*960 gal/day*		
Irrigation system	*45 gal/min*	*120 min/day*	*5400 gal/day*		

Figure 4.6 Water waste elimination worksheet. Current state: overall building.

To measure the amount of water discharged—that is, the water that goes into the drain or directly into the sewer—we will make an assumption. The assumption is that unless water is retained or used up in the process, discharge is equal to consumption. Water could be retained if your final product contains water, and it could be used up if it evaporates during processing. If this is the case, subtract the amount retained or used up to get your discharge. To determine the amount, ask the materials manager or operations manager how much water is retained or used up. Once you have determined this number, fill in the worksheet, as shown in Figure 4.7.

At Greanco, none of the activities retained water, so the team determined that the discharge is equal to the consumption. They also assumed that all sinks and toilets were discharging what they consumed.

Step 3: Measure Toxicity of Water Discharged in Value Stream Activities

Measuring toxicity is an important step for dealing with water waste because it associates environmental impact with the use of water. Due to the

Water Waste Elimination Worksheet					
Activity or Area:		***Value Stream Activity 1: Forming***			
Current State					
Identify	**Measure**				
Item	Flow Rate	Usage	Consumpton	Discharge	Toxicity
Sprayer 1	*3 gal/min*	*60 mins/day*	*180 gal/day*	*180 ga/day*	
Forming machine	*0.25 gal/min*	*420 mins/day*	*105 gal/day*	*105 gal/day*	
Wash basin	*1 gal/min*	*90 mins/day*	*90 gal/day*	*90 gal/day*	

Figure 4.7 Water waste elimination worksheet. Current state: value stream activity 1.

complexity of testing water for toxins, it is not suggested that you try to perform this step yourself unless you have the in-house expertise to do it. The suggested course of action for completing this step is to have your water tested by a third-party laboratory. A simple Internet search for "water testing in [your area]" will yield a number of organizations that can do this for you quickly. The testing is inexpensive, and chances are that if you are discharging contaminated water, you are required to be reporting it and have already had it tested. Check with the plant manager, EHS (environmental health and safety) representative, or whoever is responsible for this in your organization for any existing test results. Once you have this information, note it on your worksheet, as shown in Figure 4.8. Performing this step will aid you in minimizing the toxicity of your water, because we know that you cannot fix something if you aren't measuring it.

The Greanco green team tested the discharged water by providing sample bottles of the discharged water to a lab. When the results came back, team members found out that oil, which was on the steel when it came to them, was being washed off by the sprayer and discharged into the drain with the water. The water coming out of the forming machine contained tiny pieces of steel that broke off in the forming process and were going down the drain with the water. They decided not to have the washbasin tested, as they already knew that oil and grease were being washed off the tools and other items and being sent down the drain. They noted the toxicity of the discharged water, as shown in Figure 4.8.

Water Waste Elimination Worksheet					
Activity or Area:	***Value Stream Activity 1: Forming***				
Current State					
Identify	**Measure**				
Item	Flow Rate	Usage	Consumpton	Discharge	Toxicity
Sprayer 1	*3 gal/min*	*60 mins/day*	*180 gal/day*	*180 ga/day*	*Oil 1 ml/gal*
Forming machine	*0.25 gal/min*	*420 mins/day*	*105 gal/day*	*105 gal/day*	*Steel 1 mg/gal*
Wash basin	*1 gal/min*	*90 mins/day*	*90 gal/day*	*90 gal/day*	*Oil/grease*

Figure 4.8 Water waste elimination worksheet. Current state: value stream activity 1.

Step 4: Minimize the Consumption of Water

Now that we have identified and measured the source and use of water in each of the value stream activities as well as the overall building, we can begin to develop ways to minimize this use of water. This is an important step because it produces immediate environmental and economic savings and also sets you up to start moving toward the final two steps, which are rainwater harvesting and reuse. To start this step and any of the other minimize steps, choose the items that are the worst offenders and focus on cutting those in half, and then move on to the next biggest offenders. There are typically two ways to minimize water consumption before moving on to rainwater harvesting and water reuse: (1) using water-efficient technologies that allow you to do the same work, but use less water, and (2) practicing conservation techniques, such as turning off the tap instead of letting it run.

Take a look at the biggest consumers of water from the activities both inside and outside your value stream, as noted on your worksheets, and ask a couple of key questions. The first question is: *Are there water-efficient technologies that could replace existing hardware to save on water consumption?* If you don't know the answer to that question, call the device manufacturer and ask whether they have any models that use less water, or do an Internet search for "water efficient [name of device]." Once you have identified a water-efficient device you can use, you need to determine how much less water will be used. This is simple: If Item A uses 5 gallons per minute and the more efficient Item B uses 3 gallons per minute, you know you will save 2 gallons per minute. Multiply 2 gallons per minute saved by the number of minutes it is being used per day, week, or month. You get the idea; you are trying to figure out how much water you would save by using this device. Many times, the people selling these devices have calculators to help you, or they will even make the calculations for you in hopes that you will buy the product. Take advantage of these free services, and don't be afraid to ask for them either.

The next question you want to ask is: *Whether or not we are using the most water-efficient device, how could the water consumption of a given activity be reduced?* Look for any leaks in the devices that are delivering or dispensing the water. One small drip every second results in over 50 gallons of wasted water per year. Also, see whether water is being dispensed when it is not required, similar to the idea of leaving the water running while brushing your teeth. If you are looking for more ways to conserve, you may want to consult your local plumber or the utility delivering the water to you and ask them for ways to reduce the consumption of water, or do some research yourself by looking for others who have faced the same challenge.

What's important to remember here is that there are some easy ways to minimize water usage and quickly get some savings: replacing old, inefficient fixtures with efficient ones; ensuring there are no leaks; and so on. Then, there are solutions that will take a little more time and effort. In either case, however, remember that unless what you are doing is extremely unique and rare, someone has probably already figured out a solution, so piggyback on that solution and implement it in your facility. Once you have decided on appropriate initiatives, note the actions you plan to take and the estimated savings in the Future State section of the Water Waste Elimination Worksheet, as shown in Figure 4.9.

Next, develop an action plan with a deadline to implement the initiatives. You will also want to look at ways to finance these initiatives, as it may make sense to lease new fixtures or equipment to produce immediate positive cash flow. In addition, look for incentives that may be available to those who are taking this step to help the environment. Call your local, state, or federal government and ask whether there are any incentive programs available, but make sure you do it *before* you implement anything. Next, make sure you advertise your initiatives to your employees, customers, and suppliers to get the most mileage out of your efforts. However, don't "greenwash" by exaggerating your efforts, which can result in serious negative impacts.

Water Waste Elimination Worksheet			
Activity or Area:	***Value Stream Activity 1: Forming***		
Future State			
Minimization			
Item	Quantity	Toxicity	Savings
Sprayer 1	*Low-flow sprayer*		*60 gal/day*
Forming machine			
Wash basin			

Figure 4.9 Water waste elimination worksheet. Future state: value stream activity 1.

The Greanco green team decided to focus on the biggest consumer of water in this activity, which was the sprayer. By doing a little research and talking to a few suppliers, team members discovered a low-spray head that did just as good a job washing, but used less water. It discharged only 2 gallons per minute instead of 3 gallons, saving 1 gallon per minute, or 60 gallons per day, as noted in the worksheet in Figure 4.9.

When it came to the overall building, the Greanco green team chose to focus on the largest offender, in this case the irrigation system. Because the grass and foliage never appeared dry or brown, they decided to test putting the sprinklers on only at night, which would save one hour of usage or 2,700 gallons of water a day. They noted this in their worksheet, as shown in Figure 4.10. If this works, with no effect on the foliage, they plan to test irrigating only once every other day and see how that goes.

Step 5: Minimize the Toxicity of Water Discharge

Tackling the challenge of minimizing the toxicity of water discharge will bring you significant savings and have a huge impact on the environment. Savings in this area can have very quick paybacks and very positive impacts on reputation, customer loyalty and attraction, and employee loyalty.

Water Waste Elimination Worksheet			
Activity or Area:	***Overall Building***		
Future State			
Minimization			
Item	Quantity	Toxicity	Savings
Tooilets (10)			
Sinks (8)			
Irrigation system	*Changed timer*		*2,700 gal/day*

Figure 4.10 Water waste elimination worksheet. Future state: overall building.

Because you already know the toxicity of your water and have minimized the amount being contaminated in the preceding step, start to come up with solutions for cleaning it up, so that eventually you can reuse it and avoid the fees associated with discharging contaminated water. These fees can be large, and not having to pay them will bring significant savings year after year.

A good way to start this process is to figure out what your fines, fees, or penalties are for discharging contaminated water. Ask your plant manager, EHS representative, or controller for this information. Once you know how much you are paying and the agency that is charging you, talk with them and see what would have the greatest impact in reducing your fines or levies. If you eliminated one type of toxin, would it have the greatest impact? What if you lowered concentrations across the board? This step is key because you will want to focus your efforts on what is going to have the greatest impact on the bottom line. If you are not paying any extra fees or you are already cleaning the water yourself, focus on the largest offender. You can easily determine this by looking at your toxicity test results. Is the largest toxin in your water lead, oil, mercury, or something else? Either way, you will now have a targeted toxin to get out of your water, either the largest offender or the one that will give you the biggest savings. Once you have identified this toxin, focus on finding ways to get it out of your water.

First, try to eliminate the source of the toxin. If you are using a contaminating material or chemical, is there a substitute you could use? If you don't know, talk to the supplier of the chemical, or research what others have done. Chances are very good that you are not the only one who has faced this challenge. Search the Internet for "substitutes for [toxin x]." Contact your

CASE STUDY: BULL INFORMATION SYSTEMS, INC.

Bull Information Systems, Inc., is a provider of information technology systems and manufacturer of printed circuit boards. Due to the highly competitive nature of their industry and their commitment to sustainability, they have undertaken many water conservation initiatives to cut costs and reduce their impact on the environment. One of their many initiatives involved minimizing water use through simple conservation practices. Their manufacturing process included a plating process line in which water was a key ingredient. In order to minimize the amount of water they were using in this line, they installed automatic shut-off valves on some of the machines and manually shut off the water flow on other machines so that water flow stopped when the machine was not running or product was not going through. The total cost of the initiative was approximately $3,000, and annual water savings is 20,000,000 gallons. At an average rate of 80 cents for 500 gallons of water, this equates to annual savings of $32,000 without even incorporating the savings on municipal wastewater fees.

local environmental organizations. They will be glad to help if they can; that's what they are there for.

Only after you have exhausted ways to eliminate the source of the toxin should you look into cleaning the toxin from the water so that the water can eventually be used again. If you can avoid cleaning it in the first place, you will save money by not having to purchase the ability to clean it. See whether there is a company that specializes in doing this. You can easily find out by doing a simple Web search on "removing [toxin x] from water" or "treating [toxin x] in water." Once you have a solution, fill in the worksheet, as shown in Figure 4.11. Again, if you are having trouble finding a solution, speak to a government or environmental organization that can point you in the right direction. These organizations exist to help—and want to help—and tend to be underutilized by businesses. The people working at these organizations have many of the answers you are looking for, so don't waste your time reinventing the wheel. They will be happy to see that your organization is taking action toward combating environmental destruction and will probably give you good press. However, it is important to note here that I do *not* suggest approaching environmental enforcement agencies for help. Your good intention to improve may end up in a sour outcome. If you

Water Waste Elimination Worksheet			
Activity or Area:	***Value Stream Activity 1: Forming***		
Future State			
Minimization			
Item	Quantity	Toxicity	Savings
Sprayer 1	*Low-flow sprayer*		*60 gal/day*
Forming machine		*Magnetic filter*	*1 mg/day*
Wash basin			

Figure 4.11 Water waste elimination worksheet. Future state: value stream activity 1.

are unsure whether a particular group is an enforcement agency, do a little research to find out. Likewise, before you approach anyone for help, it is prudent to do appropriate research into who they are, not only to safeguard you from potentially airing your dirty laundry to the wrong people, but also to ensure that you get reliable advice.

Greanco wanted to stop sending steel shavings from the forming machines down the drain. Through a little research, they found out that a company with a similar problem filtered discharged water through a magnetic filter to catch all the steel shavings. It takes only a minute or two to clean the filter once a week, and they recycle the steel shavings caught in the filter. They noted their solution on the worksheet as shown in Figure 4.11.

Step 6: Self-Harvest Rainwater

Rainwater harvesting is a great way to cut down on the costs associated with purchasing water, which will no doubt continue to increase as the laws of supply and demand begin to have a greater effect on the price of this limited natural resource. It also has a positive impact on the environment, as it eliminates the water treatment process and its associated negative environmental impacts, and it eliminates the need to pull water out of aquifers, lakes, rivers, etc. For this reason, it is encouraged that you do this step in tandem with your water minimization and cleansing activities to bridge the gap to the end goal of continual reuse. The harvesting of rainwater is gaining in popularity as people start to see the associated benefits. The upfront costs are relatively low, with the largest cost being that of the tank to hold the rainwater. It is also relatively easy to do, and the operating costs are minimal. Although the steps for setting up a rainwater harvesting system vary depending on variables such as whether you need to filter the water and the size of the system, here are some general guidelines you can follow to get an idea of the potential at your site for rainwater harvesting.

First, contact your local municipality to see whether you are allowed to collect rainwater. Some municipalities, although very few, see it as stealing from others' ability to retrieve water from the local watershed. You also want to find out whether there are any incentives or help available in undertaking this project, as many do provide incentives and tax breaks. Assuming you are allowed to harvest rainwater, determine the demand for and supply of rainwater. Obviously, it is ideal if the supply is larger than the demand; if not, you will have to pick and choose which items or processes will be supplied with rainwater and which ones won't.

To determine the supply, first determine the surface area or footprint of your roof. If you do not know, ask the facilities manager, landlord, or plant manager. Otherwise, get a rough estimate of the length and width of the roof and multiply them together to get the square footage. This is the footprint. Once you have the footprint, use the general guide that one square foot of roof (collection surface) provides 0.62 gallons of water per one inch of rain. So, if you have 50,000 square feet of roof surface, you would be able to collect 31,000 gallons of water per inch of rain. Next, determine the average or monthly rainfall amounts in your area. You can get this from a number of sources, including local and state governments and rainwater collection associations. Of course, you can always use the Web to search for "monthly rainfall amount in [your city]," and a number of resources will pop up. Next, multiply the average amount of rainfall by the amount that your roof footprint will allow you to collect. For example, if the average rainfall in your area is one inch per month, you could collect 31,000 gallons per month. The collection amount is a guideline that varies depending on the surface materials of the roof, but it can be used to give you a good, quick estimation of your potential supply of rainwater.

Water Waste Elimination Worksheet					
Activity or Area:	***Value Stream Activity 1: Forming***				
Green State					
Elimination					
Item	Rainwater (*insert harvest amount*)			Re-use	
	Filtered?	Supply RW?	Savings	Grade	Use
Sprayer 1	*No*	*No*			
Forming machine	*No*	*No*			
Wash basin	*No*	*No*			

Figure 4.12 Water waste elimination worksheet. Green state: value stream activity 1.

$$\text{roof surface area} \times 0.62 = \text{rainwater supply per inch of rain}$$

$$\text{roof surface area} \times 0.62 \times \text{average rainfall (month/week/day)} = \text{rainwater supply (month/week/day)}$$

Determining the demand for rainwater is even easier. Just look at your worksheet to determine the overall demand figures you have already calculated in the Measuring step, and add them all up. If the demand exceeds the potential rainwater supply you have calculated, you will need to choose which items or activities you want to supply with rainwater. Next, determine whether the water needs to be filtered; in other words, determine whether it needs to be potable or drinking water. Fill in the answer to this question in the Green State section of the Water Waste Elimination Worksheet for each item you have identified as a consumer of water, as shown in Figure 4.12. From this, you will know what amount of rainwater needs to be filtered. Of course, you could choose to supply only those that do not require filtered water and forgo the expense of the filter for now. Note what items you will supply with rainwater by indicating it in the "Supply RW?" column of your worksheet.

As you can see with Greanco, none of the applications for forming require that the water be filtered; rainwater will work just fine, and it has noted this in its worksheet, as shown in Figure 4.12.

For the building items that are consuming water, only the sink requires that the water be filtered. This water is used to wash fruits and vegetables, to wash dishes, and as drinking water, so it will need to be filtered if it is to be supplied by rainwater. Since team members do not have enough rainwater to meet the current water demands of all activities and the overall building, they have decided to use all rainwater collected to feed the irrigation system. They have a 50,000-square-foot roof footprint, which can collect 31,000 gallons of water per month. This will supply approximately 11 days of water (31,000/2,700 per day) per month to the irrigation system. Their goal is to minimize water consumed by the irrigation system to the point where they can feed it every day from rainwater and not have to use or pay for potable water. They have noted this on their worksheet, as shown in Figure 4.13.

Once you know the supply of and the demand for rainwater, use this information to determine the storage capacity requirements of the container that will be used to hold the water. The container needs to be able to handle both peaks in supply (for instance, during thunderstorms) and peaks in demand, so take that into consideration when coming up with the size of the tank you will need. The tank comes in many different shapes and forms and costs anywhere from 50 cents a gallon for a standard fiberglass tank to $4 a gallon for a high-end welded steel tank. The cost for the tank is going to be the single largest cost of the entire system.

Next, choose a filter (if you need one) based on your needs. There are many types of filters, such as UV lights, chlorine, and charcoal. The cost of the filter varies according to type, amount to be filtered, and other factors.

Water Waste Elimination Worksheet					
Activity or Area:	***Overall Building***				
Green State					
Elimination					
Item	Rainwater (*insert harvest amount*)			Re-use	
	Filtered?	Supply RW?	Savings	Grade	Use
Toilets (10)	*No*	*No*			
Sinks (8)	*Yes*	*No*			
Irrigation system	*No*	*Yes*	*31,000 gal/mth*		

Figure 4.13 Water waste elimination worksheet. Green state: overall building.

Your best bet is to call in a water filter specialist to discuss what filter best suits your needs and the costs of that filter. Costs vary from a couple hundred to a couple thousand dollars. You should have no trouble locating a water filter specialist via the Internet or even the yellow pages.

Next, you need a pump (make sure it has an energy efficient motor!). The pump size will be based on the amount of water to be pumped. Pumps are fairly inexpensive, ranging anywhere from a couple hundred to a couple thousand, based on size. For information on sizing pumps, see the reference located at the end of this section.

Finally, you will need to tie this into your existing plumbing system so that you can use the water you harvest. This can be done by contacting your local plumber.

So, let's recap the steps for harvesting rainwater:

- Contact local authority to get permission and incentives information
- Determine supply
- Determine demand
- Determine holding tank size
- Evaluate filter requirements
- Evaluate pump requirements
- Tie into existing plumbing system

This is by no means a technical manual for the design and installation of rainwater harvesting systems. Rather it is intended to provide you the basic "how to" of harvesting rainwater and provide the ability to get a rough estimate of the cost and payback involved so you can determine the priority of the project relative to other waste-elimination projects. To calculate the savings you would receive by harvesting rainwater, multiply your per-unit charge for water times the amount of rainwater you would harvest. (You can find your per-unit charge for water on your water bill; it averages 80 cents per 500 gallons but will vary from region to region.) In the example above, 31,000 gallons of water harvested at 80 cents per 500 gallons is $49.60 per month or $595 per year savings. Knowing the cost of the tank, filter, and pump will then allow you to calculate a simple payback. Typical paybacks for commercial- and industrial-size buildings are between 3 and 5 years, depending on precipitation levels. For more detailed information on rainwater system design, tank sizing, and filters, the Texas Water Development Board provides an excellent manual with all the details.

RAINWATER HARVESTING IN PRACTICE

A commercial office building in Manchester with a roof surface of 3,200 square meters recently installed a rainwater harvesting system. The Manchester area receives 806 mm per year of rainfall, from which it is able to harvest 2,323,000 liters of rainwater. The building's manager decided to use this rainwater to supply the water needed to flush toilets for 550 personnel who occupied the building. The capital cost of the system was $12,000, but they are saving $4,000 annually, for a simple payback of three years on the capital cost of the system.

You can find the link to this and other rainwater harvesting resources in Appendix C.

Of course, you could also simply put a barrel or a container at the end of your eaves troughs to collect rainwater. Many people do this; it works pretty well and costs little to no money. Once you have collected water in the barrel, fill watering cans for irrigation of office plants or use it however you see fit.

Step 7: Transition toward the Continual Reuse of Water

Having minimized the use of water throughout the value stream and in the overall building, you are left with the consumption of water you really need to perform the activities in your value stream and run your business. By cleaning the water you discharge, you are able to use it again and avoid some of the fees and levies associated with discharging contaminated water. By harvesting rainwater, you have minimized the amount of water you are paying for. From here, focus on reusing the water you are paying the local provider to supply, and then, if you are really green, you can focus on reusing even the rainwater.

At first, the reuse of water seems rather complicated, but it's definitely not rocket science. It involves the simple task of capturing and storing discharged water, and then feeding the stored water back into your existing plumbing system. Start this step by grading the discharge of each item you have identified. Look at each item you identified in your Water Waste Elimination Worksheet. Is the water discharged by that item clean water (that is, suitable for drinking), gray water (not suitable for drinking but suitable for other use), or black water (must be treated by a professional-grade treatment

Water Waste Elimination Worksheet					
Activity or Area:	***Value Stream Activity 1: Forming***				
Green State					
Elimination					
Item	Rainwater (*insert harvest amount*)			Re-use	
	Filtered?	Supply RW?	Savings	Grade	Use
Sprayer 1	*No*	*No*		*Black*	
Forming machine	*No*	*No*		*Grey*	*Sprayer 1*
Wash basin	*No*	*No*		*Black*	

Figure 4.14 Water waste elimination worksheet. Green state: value stream activity 1.

system, for example, water flushed down the toilets)? Note the grade of each type of discharged water on your worksheet, as shown in Figure 4.14.

The Greanco green team has assigned a grade to the discharge of water by each item identified. Team members know that there is oil in the discharge of the sprayer water and the washbasin, so they have labeled it as black. However, the water being discharged from the forming machine is now being filtered and, while not drinkable, it could be used again; they label it as gray, identifying it as a potential source of water that could be reused.

The Greanco green team identified the toilet discharge as black water, but the sink and irrigation discharge as gray water because, although not potable, it could be reused.

Next, determine where you want to use that gray or clean water. Your choices are already narrowed down for you, as you can consider only those activities that can handle unfiltered rainwater. Starting with those items, look for where you could use the gray water to feed another consumer of water. Once you have decided, note it on the worksheet as Greanco did in Figure 4.15, then call in your plumber and get it done.

As you can see, this is not very difficult when you have a systematic process in place. In fact, this is already being practiced in a number of places with great success. Completing the final step is where the benefits of focusing on water as a waste and following a simple process will have a huge

Water Waste Elimination Worksheet					
Activity or Area:	***Overall Building***				
Green State					
Elimination					
Item	Rainwater (*insert harvest amount*)			Re-use	
	Filtered?	Supply RW?	Savings	Grade	Use
Toilets (10)	*No*	*No*		*Black*	
Sinks (8)	*Yes*	*No*		*Grey*	*Toilets*
Irrigation system	*No*	*Yes*	*31,000 gal/mth*	*Grey*	

Figure 4.15 Water waste elimination worksheet. Green state: overall building.

impact. Tackling water as a waste is a relatively simple way to reduce your environmental impact dramatically and, with fairly quick paybacks, have a positive impact on the bottom line. Remember that these solutions can often be financed in the same manner as energy solutions, by paying out of your savings (performance contracting), thereby bringing immediate positive cash flow to your organization. Failure to at least look at your value stream from the perspective of the environment with water as a focus is similar to flushing dollar bills down the toilet. It's throwing away money and hurting the environment. In almost every case, doing the right thing for the environment brings financial benefits, and vice versa.

CASE STUDY: CARLOVER'S CARWASH, LTD.

Carlover's Carwash, Ltd., operates 87 carwash stations throughout Australia. The owners are committed to sustainable development and, with water being the main tool they use for their business, they understand the importance of conserving it. Their preexisting process discharged all water used to wash cars into the sewers. Their new process now uses only biodegradable cleaners and employs a water recovery system so that the discharged water can be reused. This initiative allows them to recover more than 80 percent of the discharged water and saves them over—are you ready?—1 billion (with a "b") liters of water per year. If you are a competitor of theirs and not reusing water, you might want to think about it.

Chapter 5

The Third Green Waste: Materials

In order to introduce the green waste of materials, let's revisit the tree concept discussed earlier. The first time I was introduced to this idea, I was stunned. It made so much sense; it was so simple; and all it takes to get it is a simple shift of thinking, which we have already explored and will continue to explore throughout this book. A tree naturally and effortlessly pulls together a number of resources and activities to produce a product (leaves) and a service (cleaning the air and purifying water) that in turn feed its own growth. Nothing is wasted; there are no negative impacts on the environment; and in fact there are positive impacts.

Imagine a business that followed this simple principle. Imagine you could provide the product you make now, but that after its life was over (just as when a leaf falls off a tree), the product would come back into your business, helping you to grow and produce more product instead of going to the landfill or being down-cycled into something of lesser value. Think about this for a second: Say you were making a computer or a car designed in such a way that you could take it back, make some alterations or refurbish it to meet new requirements, and then turn around and offer it as a new product. What you could not reuse would go back into the earth to help enrich it.

This concept of reuse is not a new idea and is well documented and explained in the best-selling book, *Cradle to Cradle*, by Bill McDonough and Michael Braungart. Consider the concept as a solution for eliminating the green waste of materials. By eliminating materials waste, you will eliminate or dramatically reduce the need to pull virgin materials to make a product, and thereby have a tremendous impact on your double bottom line—that is, the environmental and financial bottom line. There will be no more environmental impact from pulling those materials into your value stream, and there will be little or no cost to provide the raw materials into your value stream. Your value-adding activity becomes one of tweaking an existing material, as opposed to harvesting a virgin material.

Some people doubt the practicality of this concept, but it is already being done. The real waste of materials is the negative environmental and financial impacts of pulling virgin materials and not being able to reuse them. Therefore, the end goal in tackling this waste is to mimic the activities of a tree and be able to return all your outputs either back into your value stream, allowing you to produce more product, or back into the earth as a nutrient. This is also referred to as the "cradle to cradle" cycle, closed-loop cycle, or reuse. Of course this is a large task and takes time, which is why there are intermediate steps that will provide immediate benefits on the road to the end goal of total reuse. Let's look at the steps in the process for the overall elimination of the green waste of materials:

Step 1: Identify the input and output of materials in each activity of your value stream as well as in the overall building.
Step 2: Measure the recycled/recyclable and compostable content of each material input and output.
Step 3: Classify each material input and output as a biological nutrient, a technical nutrient, or neither.
Step 4: Assess materials according to their impact on environment and society.
Step 5: Phase out materials with negative environmental impact.

INTENDING TO REUSE MATERIALS

Once you have the intention to reuse materials, you start to look at things differently and ask yourself how your products could be reused. This sparks great ideas and innovations, which can produce dramatic results. In order to be successful in achieving the end goal of this waste, you must have the intention to do it at the start, at the design stage. Without this intention, it becomes very difficult to achieve the end goal, as your products have not been designed to be reused.

Step 6: Minimize the materials used.
Step 7: Move towards 100 percent recycled/recyclable or compostable material inputs and outputs.
Step 8: Move toward 100 percent reusability, either as a technical or biological nutrient of all material outputs.

Step 1: Identify the Input and Output of Materials in Your Value Stream

In this first step, all we want to do is identify the input and output of materials for each activity of the value stream. (Note that if water is one of the inputs of your value stream, then you have already addressed it in tackling water as a waste, and there is no need to identify it here.) This task seems daunting at first due to the large amount of materials that are used, but once broken down, it becomes simple, so let's get started. There are four levels of identification for materials:

1. Identify the input of parts and their amounts that are going into the product at each activity of the value stream (nuts, bolts, enclosures).
2. Identify the output of the materials coming out of the activity (you will need this for later steps).
3. Identify the material makeup of the input parts (e.g., *steel* nuts and bolts, *plastic* enclosures).
4. Identify the chemical makeup of the input parts, down to 100 ppm, for the steel, plastic, or whatever else the material might be.

On the first pass, you may choose to do only Steps 1, 2, and 3, saving Step 4 for later. You can dive into the materials that make up the parts you are using fairly easily and quickly, but getting down to the chemical composition (the 100-ppm level of the materials) is more intense and time consuming. However, for the greatest benefits, it is encouraged that you try to move toward identifying materials down to the 100-ppm level. This is the requirement for cradle-to-cradle certification, if you should choose to seek that certification.

Here are a couple of approaches to identifying the parts that are inputs at each activity. The first and easiest is to obtain a bill of materials (BOM) for the product(s) in the value stream you are focusing on; the BOM lists all the parts and materials that go in and their amounts. A detailed BOM will

Materials Waste Elimination Worksheet									
Activity or Area:			***Value Stream Activity 1: Forming***						
Current State									
Identify	**Measure**								
Input		Output	Material Makeup		Classify				
Item & Qty	Material Makeup	Item	Input	Output	Tn	Bn	LF	Assess	
Steel sheets (100 per day)		*Frame*							

Figure 5.1 Materials waste elimination worksheet. Current state: value stream activity 1.

break down what materials go in at each activity of the value stream. To find out whether you have a BOM, ask the materials or purchasing manager, the operations or plant manager, or one of the design engineers. If you don't have a BOM, there are a few ways to tackle this: Go to each activity in the value stream, watch what goes in and how much, and ask the workers what and how much goes in. Another way is to take the WIP (work in process) piece after the activity and examine what and how much it contains. You can also ask the purchasing manager, operations manager, or plant manager what and how much is going in at that stage. In any case, you should be able to identify a majority, if not all, the parts and how much is input at each stage. Once you have this information, fill in the Current State section of the Materials Waste Elimination Worksheet, as shown in Figure 5.1.

In later steps, you will need to make sure that you haven't altered the recyclability or compostability of the material inputs by putting them through your value stream activity. Therefore, note the output of materials at each activity as well. This is the subassembly or work-in-process piece that will move on to the next activity. Note the output on the worksheet, as shown in Figure 5.1.

The Greanco green team went to Activity 1 (forming) and looked for materials going into the process. All the team members could see were sheets of steel going into the forming machine. To ensure this was the only input, they asked the workers in the area and confirmed that the only material going in at this stage was sheets of steel. They identified this input and the amount in their worksheet, as shown in Figure 5.1. They also noted that the material output at that stage was a work-in-process piece, a partially completed frame.

The next step is to identify the materials that these parts are made of. For example, if you identified a screw that went in at the fist stage, note what type of material that screw is made from: stainless steel? cold-rolled steel that is zinc plated? fiberglass? If you identified a plastic piece going in, what type of plastic is it? Polyurethane or polypropylene? You get the idea. You can do this fairly easily in a few different ways. Sometimes, the item itself or the packaging will indicate the material makeup of the item. If not, you can usually find out with one simple call to your supplier or the manufacturer of the item. Make sure you note the percentage of the part that the material makes up. For example, some items may contain two different types of steel or plastic, so list each steel or plastic as a percentage. You can get this information from your supplier/manufacturer as well. Once you have it, fill it into your worksheet, as shown in Figure 5.2.

Materials Waste Elimination Worksheet								
Activity or Area:		***Value Stream Activity 1: Forming***						
Current State								
Identify	**Measure**							
Input		Output	Material Makeup		Classify			Assess
Item & Qty	Material Makeup	Item	Input	Output	Tn	Bn	LF	
Steel sheets (100 per day)	*100% Cold rolled*	*Frame*						

Figure 5.2 Materials waste elimination worksheet. Current state: value stream activity 1.

Having identified sheets of steel as the only material being used in this first activity, the Greanco green team next wanted to find out what kind of steel it was. They first asked the workers operating the machine, and they did not know, so they turned to the purchasing manager who was responsible for buying the sheets of steel. She informed the green team that it was cold-rolled steel, so they noted that on their worksheet, as shown in Figure 5.2.

To determine the chemical composition of the material input down to the 100-ppm level is a little more difficult. There are a couple of ways you can do this. First, speak with your suppliers and the manufacturer to see whether they have this information. You can also use the Web to search for the "chemical composition or makeup of [material x]." Another way to do this is to send the material to a lab for testing. Finally, you can engage in the cradle-to-cradle certification process; determining this information is part of that process.

To identify the materials being used at the building level is a little more abstract. Obviously, it would be a tedious exercise to identify every single item you use at a building level, so focus on the more obvious materials such as office supplies (copy paper), janitorial and maintenance supplies (garbage bags and cleaning agents), kitchen and lunchroom supplies, and so on. You can gather information about these items by speaking with the office manager, the janitorial/maintenance staff, or the people who are responsible for purchasing those items. Also, take a walk around the office and lunchrooms and note what you see. Once you have identified the main materials used to support the value stream in the overall building, note them on your worksheet, as shown in Figure 5.3. (Note that you may also choose to note the quantities of materials as you did with the value stream activities. This can be difficult with some items, so just do your best. It may be easier to estimate over longer periods of time, such as per month or per quarter.)

When it came to the building-level use of materials, the Greanco green team identified a number of items, including the paper used to print and copy documents, folders used in accounting and for filing customer orders, and envelopes used to send documents

Materials Waste Elimination Worksheet								
Activity or Area:		***Value Stream Activity 1: Forming***						
Current State								
Identify	**Measure**							
Input		Output	Material Makeup		Classify			
Item & Qty	Material Makeup	Item	Input	Output	Tn	Bn	LF	Assess
Copy paper (75 pkgs/mth)								
Folders, etc. (3 boxes/mth)								
Plastic cutlery (1 box of each/ mth)								
Cleaning supplies								
Toiletries								

Figure 5.3 Materials waste elimination worksheet. Current state: overall building.

such as invoices. They also identified plastic cutlery, plates, cups, and so on being used in the kitchen. Finally, they identified that they were using janitorial supplies and toiletries such as paper towels and toilet paper. They noted each of these items on their worksheet, as shown in Figure 5.3.

When it came to the building-level material inputs, the Greanco green team looked at the packaging to determine the makeup of the materials or ingredients. The copy paper was standard copy paper, and the folders were a thicker cardboard paper. The plastic cutlery and plates were stamped with a "5" recycling symbol. The team checked and found out that 5 signifies polypropylene. When it came to the cleaning supplies, they looked at the ingredients, including ammonia. The green team noted the materials or ingredients that made up each of the items they identified on their worksheet, as shown in Figure 5.4. (See Figure 5.5 for recycling symbols and what they mean.)

Materials Waste Elimination Worksheet								
Activity or Area:		***Value Stream Activity 1: Overall Building***						
Current State								
Identify	**Measure**							
Input		Output	Material Makeup		Classify			
Item & Qty	Material Makeup	Item	Input	Output	Tn	Bn	LF	Assess
Copy paper (75 pkgs/mth)	*Paper*							
Folders, etc. (3 boxes/mth)	*Paper*							
Plastic cutlery (1 box of each/mth)	*Polypro*							
Cleaning supplies	*Ammonia*							
Toiletries	*Paper*							

Figure 5.4 Materials waste elimination worksheet. Current state: overall building.

Plastic Recycling Symbols

		Typical Products	Reclycled Products
1 PET	Polyethylene Terephthalate	Soft drinks containers Peanut butter jars	Pillow stuffing
2 HDPE	High Density Polyethylene	Milk or juice jugs Some yogurt containers Shampoo bottles	Blue Boxes Playground equipment
3 PVC	Polyvinyl Chloride	Water bottles	Floor tiles Bubble wrap Traffic cones
4 LDPE	Low Density Polyethylene	Bread and grocery bags	Plastic lumber Compost bins
5 PP	Polypropylene	Syrup and ketchup bottles	Ice scrapers
6 PS	Polystyrene	Foam cups	Egg cartons
7 OTHER	Other	Safety glasses Automotive tail lights	Outdoor signs

Figure 5.5 Plastics recycling symbols and descriptions

Step 2: Measure the Recycled/Recyclable and Compostable Content of Each Material Input and Output

Measure the Recycled and Biodegradable Material Inputs

Determining what material inputs are currently made up of recycled or biodegradable content is necessary in order to start moving toward the use of 100 percent recycled or biodegradable content (one of the later steps). If you don't know what is recycled or biodegradable in your current list of inputs, you won't know where to begin. It is important to note that in this step you are not trying to see whether the material *can be* recycled, but whether it is *made up of* recycled material. For example, all paper can be recycled, but that does not mean that all paper is made of recycled material.

There are a couple of ways to determine whether an input is made up of recycled material. Many times, the presence of recycled material will be indicated right on the material or packaging, either in wording or as one of the symbols in Figure 5.4. If it is not indicated, this does not mean that it is not recycled material. A simple call to the manufacturer or supplier will provide you with the information. You can also do an Internet search on the material, which may yield the answers. Once you determine whether the material is made up of recycled content and what the percentage is, indicate it on your worksheet, as shown in Figure 5.6.

In order to determine the recycled content of the cold-rolled steel, Greanco green team members asked their steel supplier. The supplier was not aware of the amount of recycled steel in the sheets but was able to contact the steel mill that produced the material and found out that it had a minimum of 25 percent recycled content. The team noted this on its worksheet, as shown in Figure 5.6.

Determining whether the materials used in the overall building are made up of recycled or biodegradable materials should be more straightforward. These are consumer items, and since manufacturers want to advertise how "green" their products are, this information will usually be printed right on the item or the packaging itself. Check the items you have identified for any labeling that indicates the percentage of recycled or biodegradable materials. Again, if the label does not say right on it, contact the supplier or manufacturer directly to find out. Once you have this information, enter it on the worksheet, as shown in Figure 5.7.

Materials Waste Elimination Worksheet								
Activity or Area:		***Value Stream Activity 1: Forming***						
Current State								
Identify	**Measure**							
Input		Output	Material Makeup		Classify			
Item & Qty	Material Makeup	Item	Input	Output	Tn	Bn	LF	Assess
Steel sheets (100 per day)	*100% Cold rolled*	*Frame*	*25% recycled*					

Figure 5.6 Materials waste elimination worksheet. Current state: value stream activity 1.

Materials Waste Elimination Worksheet								
Activity or Area:		***Value Stream Activity 1: Overall Building***						
Current State								
Identify	**Measure**							
Input		Output	Material Makeup		Classify			
Item & Qty	Material Makeup	Item	Input	Output	Tn	Bn	LF	Assess
Copy paper (75 pkgs/mth)	*Paper*		*60% recycled*					
Folders, etc. (3 boxes/mth)	*Paper*		*60% recycled*					
Plastic cutlery (1 box of each/mth)	*Polypro*							
Cleaning supplies	*Ammonia*							
Toiletries	*Paper*		*100% recycled*					

Figure 5.7 Materials waste elimination worksheet. Current state: overall building.

For the building-level materials at Greanco, the recycled content was printed on the containers or packaging of the materials. The paper was only 60 percent postconsumer recycled material. The plastics, along with the cleaning supplies containers, did not have any recycled content. However, the toiletries (toilet paper and paper towels) were, to the team's surprise, 100 percent postconsumer recycled content. They noted this on their worksheet, as shown in Figure 5.7.

Measure the Recyclable and Compostable Material Outputs

In the last step, we looked at the recycled and compostable/biodegradable content of material that went into a value stream activity or was being used in the overall building to support the value stream. In this step, we will look at the material outputs to see whether they can be recycled or composted. (Because the materials being used in the overall building to support the value stream are not typically changed or altered, there is no need to do this for the materials being used in the overall building.) At first glance, you might say, "If you have recycled inputs, wouldn't the material outputs be recyclable as well?" The simple answer is, "No, not necessarily." Your process may alter the makeup of the material and change its ability to be recycled or composted. A good example of this is the melting together or fusing of two metals, or the application of toxic coatings, varnish, or epoxies to compostable material, making it unsafe or impossible to compost. Also, in the last step we were looking only at whether the material was made up of recycled material, not at whether it could be recycled; it could be 100 percent virgin steel (no recycled content), but that steel is still recyclable. Remember that there are two different considerations: The first is having material input be made up of recycled or biodegradable content, and the second is being able to recycle or compost the output or finished product.

To find out whether materials can be composted or recycled, go to the activity and look at the products or materials coming out of the process. Was anything done to the material inputs that could affect their ability to be recycled or composted? If you simply form or alter the shape of the material or bolt two materials together, there's a good chance that the answer to this question is no. However, if you fused two metals together or applied toxic

Materials Waste Elimination Worksheet								
Activity or Area:		*Value Stream Activity 1: Forming*						
Current State								
Identify	Measure							
Input		Output	Material Makeup		Classify			
Item & Qty	Material Makeup	Item	Input	Output	Tn	Bn	LF	Assess
Steel sheets (100 per day)	*100% Cold rolled*	*Frame*	*25% recycled*	*100% recyclable*				

Figure 5.8 Materials waste elimination worksheet. Current state: value stream activity 1.

chemicals or something else that could changed the materials' ability to be recycled or composted, the answer to the question is probably yes. If this is the case, the material may no longer be recyclable or compostable. Either way, note the recyclability and compostability of the material output on your worksheet, as shown in Figure 5.8.

The Greanco green team reviewed their forming process and saw that during this process the shape of the steel was being changed, but no coatings were being applied and no metals were being fused or melted together. They concluded that, although only 25 percent of the steel was made of recycled content, all the steel could be recycled and, therefore, 100 percent of the outputs of this process (steel) were recyclable. They noted this on their worksheet, as shown in Figure 5.8.

Step 3: Classify Materials as Technical Nutrient, Biological Nutrient, or Neither

At this point, you have identified the materials going in and coming out, and measured how much recycled or biodegradable material makes up the inputs and how much of the outputs can be recycled or composted. Doing this allows you to choose which stream the material will follow after the end of its life—whether it will be used again to make a new product (technical nutrient), be put back into the soil (biological nutrient), or neither, which means that it will sit in the landfill forever.

A biological nutrient, as defined by McDonough Braungart Design Chemistry (an environmental consulting company), is "a material used by living organisms or cells to carry on life processes such as growth, cell division, synthesis of carbohydrates and other complex functions. Biological nutrients are often carbon based compounds that can be safely composted and returned to soil." A biological nutrient, therefore, is anything used in your product that can biodegrade and go back to enrich the earth or soil. A technical nutrient is defined as "a material of human artifice designed to circulate within technical metabolism (industrial cycles) forever." In other words, a technical nutrient is a material that will not biodegrade in the earth but can be used again in some form.

If materials follow one of these two streams or cycles to be reused, this is known as the cradle-to-cradle cycle, as it is continually reborn into something of equal or greater value (see Figures 5.9 and 5.10). If it cannot be recycled or composted, the material follows the landfill stream, meaning that it will never be used for anything good again. When a material goes to the

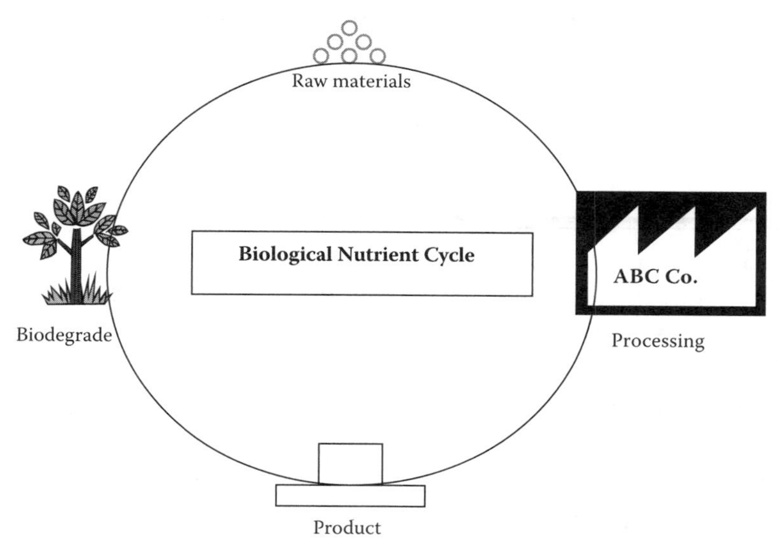

Figure 5.9 Biological nutrient cycle

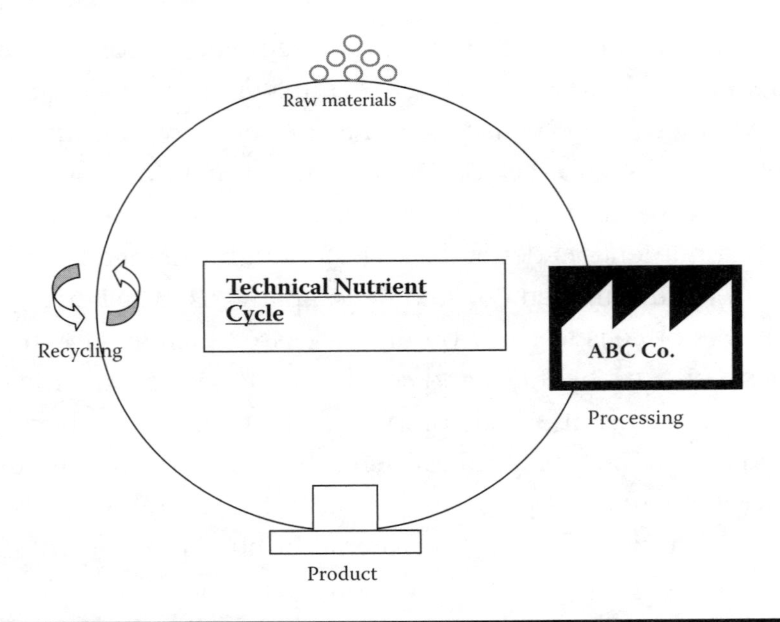

Figure 5.10 Technical nutrient cycle

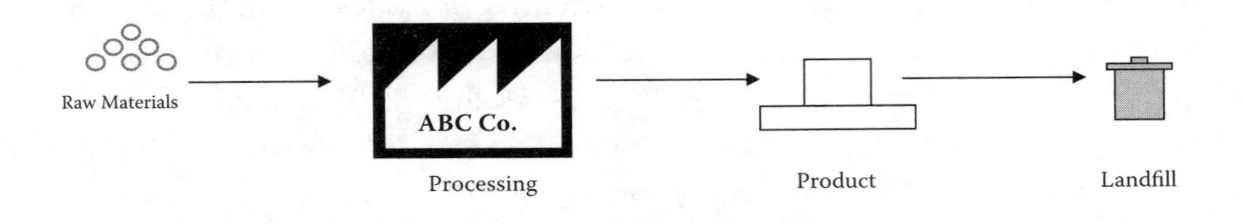

Figure 5.11 Linear material flow to landfill

landfill, it is a part of a cradle-to-grave cycle; it is born and then dies when it hits the landfill, which is a huge waste (see Figure 5.11).

The Greanco green team knew that steel cannot be composted but that it is recyclable, so it easily identified it as a technical nutrient by making a checkmark in the "Tn," technical nutrient, column on their worksheet, as shown in Figure 5.12.

At the building level, the Greanco team members reviewed each item and asked: Could this material be composted or recycled? When it came to paper, they first noted that it could be composted, but then realized that, for the most part, it doesn't get composted but, rather, is recycled, so they noted it as a technical nutrient. Plastic (polypropylene) cannot be composted but can

Materials Waste Elimination Worksheet									
Activity or Area:		***Value Stream Activity 1: Forming***							
Current State									
Identify	**Measure**								
Input		Output	Material Makeup		Classify			Assess	
Item & Qty	Material Makeup	Item	Input	Output	Tn	Bn	LF	Assess	
Steel sheets (100 per day)	*100% Cold rolled*	*Frame*	*25% recycled*	*100% recyclable*	×				

Figure 5.12 Materials waste elimination worksheet. Current state: value stream activity 1.

be recycled, so it too was labeled a technical nutrient. Because ammonia could be neither recycled nor composted, it was labeled a landfill item (see Figure 5.13).

Determining whether a material is a technical or biological nutrient or goes to landfill is quick and easy. Look at each of the materials you have listed in the worksheet as in Figure 5.8 and ask this simple question: Is the material compostable? If the answer to the question is "yes, it can be composted," it is a biological nutrient. If the answer is "no, it cannot be composted," it is either a technical nutrient or a landfill item. If the material can be recycled, it is a technical nutrient, and if it can be neither recycled nor composted, it is a landfill item. Once you have classified each of the materials as either a technical or biological nutrient or a landfill item, indicate it on your worksheet, as shown in Figure 5.12 and Figure 5.13.

Step 4: Assess the Impact on the Environment

The purpose of assessing each material based on its impact on the environment is twofold. First, if it has a negative impact, you want to phase it out, regardless of whether it can be reused as a nutrient. Second, you

Materials Waste Elimination Worksheet								
Activity or Area:		***Value Stream Activity 1: Overall Building***						
Current State								
Identify	**Measure**							
Input		Output	Material Makeup		Classify			
Item & Qty	Material Makeup	Item	Input	Output	Tn	Bn	LF	Assess
Copy paper (75 pkgs/mth)	*Paper*		*60% recycled*		×			
Folders, etc. (3 boxes/mth)	*Paper*		*60% recycled*		×			
Plastic cutlery (1 box of each/mth)	*Polypro*				×			
Cleaning supplies	*Ammonia*						×	
Toiletries	*Paper*		*100% recycled*		×			

Figure 5.13 Materials waste elimination worksheet. Current state: overall building.

want to focus on the materials that will provide the largest opportunity for environmental and economic savings in later steps.

The point of this step is to take a look at the materials you're using and try to grasp the impact they are having on the environment. If you were to throw a piece of steel or plastic into a field or lake, what would be the consequence? Would a tree, plant, fish, or other animal die or be deformed? Would it negatively impact human health?

Finding out what impact the material is having on the environment and human health can be approached in a few different ways. Research the Web by searching for "environmental impact or human health impact of [material x]" or "effect of [material x] on fish life." Approach local environmental organizations, nongovernmental organizations (NGOs), and government organizations to ask for help. The information is out there and, with a little effort, you can find it. Once you know the environmental impact a material can have on the environment, you then want to follow the cradle-to-cradle process of rating it using a color code:

- **Green:** Little or no risk to the environment or human health from using this substance.

- **Yellow:** Low to moderate risk associated with using this material. Unless you can substitute this material with one that is rated green, it is acceptable for use.
- **Red:** High impact and risk to the environment and human health from using this material. A strategy for phasing out this material and replacing it with a green or yellow material needs to be developed.
- **Gray:** Risk data is missing or incomplete. Further investigation and research is needed in order to rate the material.

Once you know the color code of each material identified, fill in your worksheet, as shown in Figure 5.14. Remember that if you identified the content at the 100-ppm level, you need to do the color coding for each of those chemicals.

The Greanco green team researched the environmental and health impacts of cold-rolled steel. Team members did not find any significant direct impact, so they decided to give cold-rolled steel a rating of yellow, as shown in Figure 5.14.

When it came to the building-level materials, the Greanco green team found out that ammonia had some major environmental and health impacts and decided to give it a rating of red; however, the other materials had little impact, and it rated them as yellow. See Figure 5.15.

Materials Waste Elimination Worksheet								
Activity or Area:	***Value Stream Activity 1: Overall Buildingv***							
Current State								
Identify	**Measure**							
Input		Output	Material Makeup		Classify			
Item & Qty	Material Makeup	Item	Input	Output	Tn	Bn	LF	Assess
Steel sheets (100 per day)	*100% Cold rolled*	*Frame*	*25% recycled*	*100% recyclable*	×			*Yellow*

Figure 5.14 Materials waste elimination worksheet. Current state: value stream activity 1.

Materials Waste Elimination Worksheet									
Activity or Area:		***Value Stream Activity 1: Overall Building***							
Current State									
Identify		**Measure**							
Input		Output	Material Makeup		Classify				
Item & Qty	Material Makeup	Item	Input	Output	Tn	Bn	LF	Assess
Copy paper (75 pkgs/mth)	*Paper*		*60% recycled*		×			*Yellow*
Folders, etc. (3 boxes/mth)	*Paper*		*60% recycled*		×			*Yellow*
Plastic cutlery (1 box of each/mth)	*Polypro*				×			*Yellow*
Cleaning supplies	*Ammonia*						×	*Red*
Toiletries	*Paper*		*100% recycled*		×			*Yellow*

Figure 5.15 Materials waste elimination worksheet. Current state: overall building.

Step 5: Phase out Negative-Impact Materials

Now that you have identified the use of materials—whether they are made up of recycled or biodegradable material; whether they can be recycled, composted, or neither; and what impact they have on the environment—you can start to take action. The first action step is to phase out those materials that have the highest negative impact on the environment, those materials identified as red. Doing this will aid you in achieving the end goal of total reuse and, in the meantime, will have a positive impact on the double bottom line. The positive impact comes from the fact that it is typically these high-risk items with negative environmental impact that have penalties or fees associated with using them: handling fees, special transportation fees, special storage requirements/fees, and so on. These items also tend to be more expensive because if they are bad for the environment, they have usually gone through a great deal of processing and have had many items added to them, increasing the costs of production. By eliminating the materials with the highest negative impacts, you will save money on these fees and penalties as well as on the buried costs of extra processing/handling. This once again proves that doing good things for the environment and society is also good for the bottom line.

Materials Waste Elimination Worksheet			
Activity or Area:	***Value Stream Activity 1: Forming***		
Future State			
Minimize			
Item	Harmful Materials (Phase Out)	Quanitity	Savings
Steel sheets	*Rated yellow, no substitute*		

Figure 5.16 Materials waste elimination worksheet. Future state: value stream activity 1.

Take a look at your worksheet and those items that have been given a red rating. Can you use a substitute for this material? To find out, do a little research by asking your suppliers, or doing a Web search on "substitute for [material x]." What have others in your industry or outside of your industry done? Could you use less of the material and more of a green or yellow material? Is there a water-based substitute for an oil-based chemical? Ask these types of questions. The key here is to develop an action plan for phasing out all the red items. Once you have the action plan or a solution, note it in the Future State section of your Materials Waste Elimination worksheet, as shown in Figure 5.16.

The Greanco green team had rated the only material in the forming process as yellow, and therefore felt no need to phase out this material based on its impact on health and the environment, as shown in Figure 5.16.

However, when it came to the building level, there was one material or ingredient rated as red that needed to be phased out. Because the ammonia in the cleaning supplies was rated as red, team members made it a policy to phase out all cleaning supplies that contained ammonia.

They contacted the cleaning company and asked for help with this initiative. After a few days, the cleaning company responded that it had found a new line of "green" cleaning supplies that contained no ammonia and would add no further cost to their cleaning bill, and furthermore that they were planning to use these supplies for all their clients. The Greanco team noted this in their worksheet, as shown in Figure 5.17.

Materials Waste Elimination Worksheet			
Activity or Area:	***Overall Building***		
Future State			
Minimize			
Item	Harmful Materials (Phase Out)	Quanitity	Savings
Copy paper			
Folders, etc.			
Plastic cutlery			
Cleaning supplies	*Use cleaners with no ammonia*		
Toiletries			

Figure 5.17 Materials waste elimination worksheet. Future state: overall building.

Step 6: Minimize Materials Usage

As with all wastes, minimizing materials use is a key step toward the end goal of reuse as a technical or biological nutrient. It allows you to minimize the effort needed to achieve the end goal, gives you immediate savings, and contributes positive benefits toward the environment. The savings result from the fact that if you use 10 percent less materials, your material cost should be 10 percent less. Of course there are exceptions, but for the most part this should be the case. If you are not getting savings from using less material, it should raise a red flag to go and see why not. The environmental benefits result from the fact that you are pulling less virgin materials, throwing less into landfills, and have less to process, all of which have positive environmental benefits. So how do we go about minimizing materials? Start by asking questions like:

- Are we putting in only the specified or designed amount of materials, or are we putting in excess?
- Do we have to put in the designed amount, or could we use less?
- Is there wastage of materials?
- Can we design differently to use less material?
- Do competitors use less material?
- How are they doing it?
- Is there someone else who may know how to do it?

Materials Waste Elimination Worksheet			
Activity or Area:	***Value Stream Activity 1: Forming***		
Future State			
Minimize			
Item	Harmful Materials (Phase Out)	Quanitity	Savings
Steel sheets	*Rated yellow, no substitute*	*Use thinner gauge*	*Material transportation cost*

Figure 5.18 Materials waste elimination worksheet. Future state: value stream activity 1.

Once you have determined some ways to minimize material usage, note them on your worksheet, as shown in Figure 5.18.

The Greanco green team noticed very little scrap or waste of steel in the forming process, and team members saw that the material going in was exactly as per the design requirements and BOM. They decided to take a look at what some of their competitors were doing and noticed that a direct competitor producing a very similar product also used cold-rolled steel in their furniture, but it seemed to be thinner than what Greanco was using. They approached their design engineers to ask why. The engineers had no answer. One of the green team members asked if they could use thinner steel, and the engineer said they could put it through some testing to see what would happen. After testing, they went ahead and changed the design to a thinner gauge steel without sacrificing any quality. In fact, their customers liked the new design better because it was lighter to move around. The result was that less material was being pulled from the ground and Greanco got a direct economic savings because the thinner gauge material was cheaper. Because it was thinner, it also took less pressure to form the steel, thereby saving energy, and less water to keep it cool. Also, because it was lighter, freight bills were a little lower.

Materials Waste Elimination Worksheet			
Activity or Area:	***Overall Building***		
Future State			
Minimize			
Item	Harmful Materials (Phase Out)	Quanitity	Savings
Copy paper		*Paper conserva-tion policy*	*Paper costs*
Folders, etc.			
Plastic cutlery			
Cleaning supplies	*Use cleaners with no ammonia*		
Toiletries			

Figure 5.19 Materials waste elimination worksheet. Future state: overall building.

When it came to the materials used to support the value stream, the Greanco green team decided to focus on reducing the amount of paper used. First, they put a policy in effect to have all printouts double-sided to save paper. Next, they put a policy into effect that people print out e-mails only if requested or absolutely required, and that all reports were no longer printed but sent electronically. Not only did this save paper, producing immediate financial savings, it also saved toner over time, and it saved people the time of printing out numerous copies of reports. They noted their focus on paper conservation policies on their worksheet, as shown in Figure 5.19.

Step 7: Move toward 100 Percent Recycled/Recyclable or Compostable Material Inputs and Outputs

Move toward 100 Percent Recycled or Biodegradable Material Inputs

By moving toward 100 percent recycled or biodegradable inputs, you accomplish a couple of things. First, you set up your process to be able to accept and use these types of inputs, helping you to accomplish the overall goal of reuse. If you can't accept these inputs to begin with, then you can't reach that end goal of total reuse. Second, you have immediate positive impacts on the environmental and economic bottom lines. Using recycled materials means you don't have the negative impacts of pulling virgin materials, and using biodegradable materials means you are using a renewable material that can then be returned to the earth or soil as a nutrient. In terms of benefits to the bottom line, recycled materials tend to be cheaper, so you also get the benefits of savings in material costs. Also, products that have this characteristic of recycled content or biodegradable content tend to be favored or required by the consumers of the product, resulting in increased demand and sales for that product, with a positive impact on the bottom line.

Because you have already measured the recycled and compostable content of each input and know those items that cannot be recycled or composted (those that go to the landfill), focus on using 100 percent recycled or compostable inputs. This often becomes a matter of simply choosing recycled material over nonrecycled material or asking your suppliers to make that choice. Other times it will involve looking around for materials that have a higher recycled content. For example, a majority of the traditional packaging materials now come in a biodegradable or recycled form, and plastics used for everything from stretch wraps to food containers are now being made from corn- or starch-based materials, as opposed to being petroleum-based.

Materials used in the overall building to support the value stream are now oftentimes available as 100 percent recycled or biodegradable. For example, 100 percent recycled paper is readily available at all big-box stores, often at the same price or less than regular, nonrecycled paper. Plastic cutlery, glasses, Styrofoam plates, and cups are now available as 100 percent biodegradable because they are made from starch instead of plastic or other materials. Note that if you choose to use biodegradable items, such as starch- or corn-based material for packaging, you need to think about what will happen to that packaging material when people are done with it. These starch- or corn-based products have a hard time decomposing in landfills: They are meant to go into a professional compost that provides the necessary environment (moisture/heat/microorganisms) to compost them, an environment the landfill doesn't provide. If you don't think these items will make it to a compost and will end up in the landfill or get mixed with recyclable items, use items that are recyclable instead of biodegradable because the net impact on the environment will be better. For janitorial products, request a company that is Green Clean certified, meaning their products and practices have been certified by a third-party auditor as environmentally friendly. Whenever possible use recycled content over organic. If everyone used organic content, there would not be enough land to support it.

Materials Waste Elimination Worksheet				
Activity or Area:	***Value Stream Activity 1: Forming***			
Green State				
Eliminate				
Item	100% Tn & Bn Inputs	100% Tn & Bn Inputs	Re-Use	Savings
Steel sheets	*In progress*	*Already 100%*		

Figure 5.20 Materials waste elimination worksheet. Green state: value stream activity 1.

One last note: If you can't find a recycled or biodegradable substitute or you are posed with a choice of more than one substitute, look for the items or materials that are either sustainably harvested or manufactured in an environmentally friendly manner. Some examples of this are FSC-certified wood that comes from a sustainably managed forest or products that come from an ISO 14001–certified plant, meaning that the environmental impact of producing those materials is continually being minimized. Once you have come up with a solution, note it in the Green State section of the Materials Waste Elimination Worksheet, as shown in Figure 5.20.

The Greanco green team members had only one material to focus on, the cold-rolled steel. Because their supplier told them it already contained some recycled materials, they approached the supplier to see whether they could supply sheets that contained more recycled content. The supplier said that at this point that it was not possible to guarantee higher levels of recycled material but that they would work on it for them. The Greanco green team members decided to let them work on this and to move their focus toward the materials in the other processes and the building. They noted on their worksheet that this was in progress so that they could monitor progress (see Figure 5.20).

For the building-level materials, Greanco green team members wanted to stay focused on paper and found that their big-box

Materials Waste Elimination Worksheet				
Activity or Area:	***Overall Building***			
Green State				
Eliminate				
Item	100% Tn & Bn Inputs	100% Tn & Bn Inputs	Re-Use	Savings
Copy paper	*Change paper*			
Folders, etc.				
Plastic cutlery				
Cleaning supplies	*Green supplies*			
Toiletries	*Green supplies*			

Figure 5.21 Materials waste elimination worksheet. Green state: overall building.

supplier of paper also offered an eco-version that was 100 percent postconsumer recycled content instead of 60 percent. The paper was actually cheaper than the current paper they were using. They spoke with their purchasing department and started a green purchasing policy; paper was the first item included in this policy. They will continue to build on this policy to implement these types of initiatives in the future. Also, they remembered that in phasing out ammonia, the cleaning service was now using green cleaning supplies. After talking with the cleaning service, they found out that all the ingredients in these supplies are biodegradable, so they noted that on their worksheet as well, shown in Figure 5.21.

Measure the Recyclable and Compostable Material Outputs

As with 100 percent recycled or biodegradable inputs, having 100 percent recyclable or biodegradable outputs will accomplish two basic things. It aids you in achieving the overall end goal of total reuse. This step is important because even if you have 100 percent biodegradable or recyclable inputs, if you do not have 100 percent biodegradable or recyclable outputs, you cannot have total reuse as either a technical or biological nutrient. Second, moving toward 100 percent recyclable or biodegradable outputs will provide immediate benefits to the double bottom line. The positive impacts on the environment are obvious. The economic benefits come directly from the fact that products with high recycled or biodegradable content are in higher demand from consumers, who are now looking for or being forced to choose products that have this characteristic. Don't forget that you will also have greater success in retaining employees and attracting the best and brightest new employees as people look to work for organizations that have considered the impact their products have on the environment.

To get started on this step, focus on only those activities or processes that took away from the recyclability or compostability of the inputs, as noted in the Measuring Recyclability and Compostability step. Dig in to determine what about the process changed the material's ability to be recycled or composted. Is it the fact that you melted or fused a recyclable and nonrecyclable material together? Did you melt something, making it no longer recyclable or compostable? Whatever it is, the challenge then becomes changing the process so that this doesn't happen. Instead of fusing, could you bolt together so the materials can be separated at the end of the product's life? For help in doing this, look to organizations that are specialists in this area, such as McDonough Braungart Design Chemistry (MBDC) or other design-for-the-environment companies. To locate such companies, do a Web search on "design for the environment services in [your area]." You can also search to see what your competitors are doing.

Step 8: Transition to 100 Percent Reuse

If you can achieve 100 percent reuse, and people already have, you will reap the greatest benefits of any step in eliminating material waste, and those benefits will astonish you. Imagine this scenario: A company builds a car, using all the raw materials that go into it, and then leases it to you. After three, four, or five years, you are finished with it and give it back. You never see the car again, and eventually, after another owner or two, the car goes to the scrap yard. A very small percentage of that car actually ends up back in any sort of product, never mind back to the car company that made it in the first place and paid for those materials. The company takes more materials out of the ground, builds another car, and you lease another car. This process is repeated over and over again.

What if, instead, a car company designed this car in such a way that, after you brought it back to them (for free by the way), they could tweak or rework 100 percent of the materials in it and make a brand new car because they had determined how to do this at the original design phase, instead of not thinking about it at all. Materials that couldn't be reused were biodegradable and went back into the earth or soil as a nutrient. You now have a car company that has pulled and paid for these precious materials once, and can reuse them over and over again forever, never having to pay for those materials again. Additionally, with many products, such as computers and cell phones, technology allows them to become smaller from generation to generation. Imagine now that you can take one car back and make two out of it, resulting in a tremendous benefit to the environment and the bottom line! Okay, you may not always be able to make two cars out of one, but how many MINI Coopers do you think could be made out of a Hummer? Anyway, you get the point, and there is actually already a car company that has made an early prototype using this concept.

Regardless of whether you are making a car, a computer, or any other product, total reuse of materials will have a tremendous direct and indirect impact on your bottom line. You will no longer be paying for materials that no longer need to be pulled out of the ground and refined, which consumes a great deal of energy resources and costs money.

The key to doing this is to follow the steps in this chapter and use 100 percent recyclable or biodegradable material inputs with 100 percent recyclable or biodegradable material outputs. Without doing this first, you can never have 100 percent reuse, because there would always be something wasted. Next, you need a design process that continually follows and supports this effort. When you design a new product or redesign an existing one, you need to do so in a way that allows you to reuse 100 percent of the technical nutrients in the same or another product, and to put the biological nutrients back into the earth. For example, if you are using a metal that is 100 percent recyclable, such as aluminum, in your product, the challenge is to create a process that can take that aluminum back from the finished product and turn it into the new product again. The next challenge is getting the product back at the end of its life. You could do this by offering an incentive to your customers to return it in exchange for a discount on the next product they buy from you. (This also helps to keep your customers coming back to you instead of going somewhere else.) By doing this, you effectively "close the loop" of your product realization process, as shown in Figure 5.22.

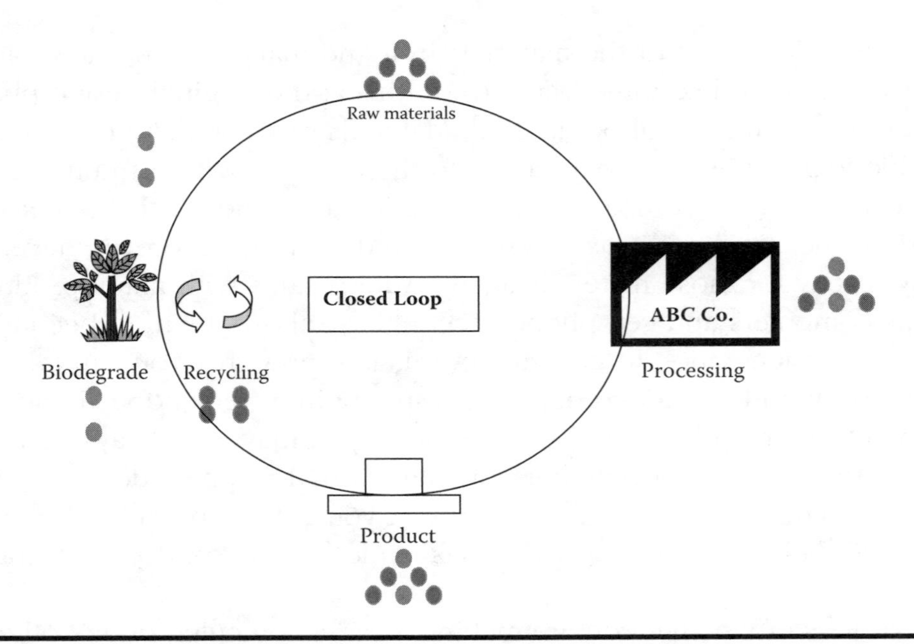

Figure 5.22 Closed loop (cradle-to-cradle) cycle

The Greanco green team members were excited about this step of the process. They knew that in the activity of forming, the output was 100 percent recyclable, and if they could figure out a way to reuse it, they would save the company a lot of money. The focus of the green team was to find a way to reuse this part. They are working with engineers to develop a solution and have noted this effort on their worksheet, as shown in Figure 5.23.

I want to close this chapter with the following thought. Hopefully it will help to shift your thinking a little more and help you move closer to the end goal of total reuse. When you consider the concept and design of a product, think about this: What do your customers really want? Do they really want a computer, or do they want to be able to surf the Internet, send e-mails, and use software? Do people really want a car, or do they just want to get from Point A to Point B, and the current concept of a car is the best way to do that right now? Do people really want a CD, or do they just want to listen to music? Many times the product you are providing people is actually a service, as in the service of word processing, or surfing the Internet, or transportation. If you look at your situation and think that it is the only way it can be done, ask yourself this: What would you do if there were no more materials

Materials Waste Elimination Worksheet				
Activity or Area:	***Value Stream Activity 1: Forming***			
Green State				
Eliminate				
Item	100% Tn & Bn Inputs	100% Tn & Bn Inputs	Re-Use	Savings
Steel sheets	*In progress*	*Already 100%*	*In engineering*	*Material cost*

Figure 5.23 Materials waste elimination worksheet. Green state: value stream activity 1.

that you could pull to make your product? Would you have to shut the doors, or is there another way you could provide what you are providing? Here's the thing: If we don't change the way we are doing things, eventually there won't be any more of the materials we need to make our products.

CASE STUDY: HEWLETT-PACKARD

Hewlett-Packard (HP) provides a great example of material reuse and needs to be commended on its overall commitment to environmental sustainability. The company is now able to take back some preowned equipment and either (a) reuse the entire piece by reformatting or refurbishing it to a buyer's specific needs or (b) reuse various parts and pieces where possible and then remarket these items as preowned. The returned items come from various sources, including customer returns and cancelled orders, products damaged during shipping, trade-in and buy-back programs, lease returns, and products used in marketing demonstrations. (HP protects customer privacy by following stringent processes for removing any personal data before resale.) This remarketing enables customers to purchase preowned equipment returned to HP, making their products accessible to more people. HP offers these remarketed products on most of their product lines, including printers, personal computers, and monitors. It has even remarketed entire data centers. Oftentimes, HP also offers a warranty for these refurbished items, showing the high quality of the remarketed product and keeping customers confident in the quality of the products HP brings to market. (For the record: HP does not yet offer remanufactured print cartridges, as they do not meet quality and reliability standards. Some other returned products are also unsuitable for remarketing by HP, including other manufacturers' equipment, and these are sold through brokers into the reuse market.)

This remarketing program now generates worldwide revenues of more than $500 million. The environmental impacts of this program are also huge, from reduced virgin material use, reduced energy use, and reduced pollution, showing that doing something good for the environment and making money go hand in hand.

Chapter 6

The Fourth Green Waste: Garbage

Garbage as a green waste comprises all the things we throw away as a result of our processes (value stream activities) or the support for those processes (overall building). This is one waste with which we are all very familiar. The idea of garbage as an environmental waste is nothing new and has come a long way over the past few years. The idea of the three Rs to reduce, reuse, and recycle our garbage is definitely not a new concept. Let's build on this traditional concept and take it a little further as we consider commercial garbage. Many organizations look at the output of garbage once it is already created, and deal with it at that point. If it can be recycled, then many of us have taken initiatives to recycle it, but we throw out the rest, forgetting or ignoring the other two Rs of reduce and reuse. We looked at the concept of reusing materials in the last chapter; in this chapter, we will explore the concept of reusing garbage as a step toward eliminating garbage. We'll get into the process for really eliminating this waste shortly, but let's take a quick look at what it means to us if we actually do it.

Picture your organization as having no "garbage" because your garbage is used either as a technical or biological nutrient or, even better, you have eliminated it completely. This means you no longer bury things in landfills to make the earth toxic for years to come; you do not contaminate people's food or water sources; and you no longer have to pay for waste removal because you reuse the "garbage" yourself, or people pay you for it because it has value to them, or, in the ultimate case, you don't have any garbage at all.

Think about the idea of creating the garbage in the first place. If you throw something out, chances are you have paid for it first and then pay again to have it thrown away. By eliminating garbage, not only do you save money on the back end, but you also save money on the front end by not having to pay for extra material that gets thrown away. Getting rid of its harmful effects and eliminating its creation in the first place will produce dramatic economic benefits, the proof of which you will see in the examples in this chapter. Because the real waste of garbage comes from both the well-known environmental impacts combined with the fact that we are paying for things we don't use—paying again to process them, and then paying again to throw them out—our end goal is the total elimination of garbage. This will eliminate the associated environmental and economic costs associated with creating and disposing of garbage and bring savings from doing a good thing for the environment.

The process for eliminating this waste follows a similar structure for eliminating the other wastes:

Step 1: Identify the creation of garbage in the activities of your value stream and the overall building.
Step 2: Measure the makeup of this garbage.
Step 3: Measure the hazardous substances in the garbage.
Step 4: Minimize the creation of garbage in your value stream activities and in the overall building and recycle what is left.
Step 5: Move toward 100 percent reusable or biodegradable garbage.
Step 6: Move toward total elimination of garbage.

INTENDING TO ELIMINATE GARBAGE

Having the intention to eliminate garbage completely will get you and others thinking about why garbage exists in the first place and how it can be eliminated. If you have the intention of eliminating garbage, you will constantly be thinking of ways to achieve that goal; without that goal and the intention to achieve it, there is a good chance you will get lost along the way.

Step 1: Identify the Creation of Garbage in Your Value Stream

As with the wastes covered in previous chapters, in this first step you are concerned only with identifying that this green waste exists. To identify garbage waste, go to the area where the first activity in your value stream is located. Look for garbage cans, bins, and bags. Once you find these containers, have a quick look inside them to make sure they actually contain garbage. If you don't see any garbage cans, ask the workers in the area or the janitor if there is any garbage being created in that area. Remember to focus on the value stream activity itself and the garbage created by that activity. If garbage is being carted away to a central location and not held in the area of the first activity of the value stream, you will still need to determine what garbage is being created in that area and carted away. Ask the workers in the area, or observe it yourself. Remember that, at this point, you are not worried about what or how much garbage there is, just that there is garbage being created at that activity. Having identified that garbage is being created, fill in the Current State section of the Garbage Waste Elimination Worksheet, as shown in Figure 6.1.

The Greanco green team members went to the first activity of their value stream and noticed two garbage cans and one larger garbage bin in the forming area. Looking inside each, they confirmed that it was garbage, and noted each container on their worksheet. They also noticed an old barrel containing hydraulic oil, and noted that on their worksheet as well (see Figure 6.1).

Garbage Waste Elimination Worksheet			
Activity or Area:	***Value Stream Activity 1: Forming***		
Current State			
Identify	**Measure**		
Item	Material	Quantity	Hazardous Substances
Garbage can 1			
Garbage can 2			
Garbage bin 1			
Oil container			

Figure 6.1 Garbage waste elimination worksheet. Current state: value stream activity 1.

Identifying garbage that is being created in the overall building is approached in the same manner. Walk around the building and look for areas that are producing garbage. Break it down by area or department, such as accounting, engineering, the kitchen, the bathrooms, copy room, specific offices, and so on. Once you have broken out specific areas to review, start to look for the garbage bins, cans, bags, and so on. If there is garbage being produced in those areas, which there probably is, note it on your worksheet, as shown in Figure 6.2. Again, if you don't see garbage cans in the immediate area, ask the people who occupy the area where they put their garbage.

The Greanco green team decided to divide the overall building into departmental areas, reflecting how the building was physically divided. One area housed operations, including reception, purchasing, and production management. In other areas were sales and marketing, engineering, and accounting. In each area, team members noted how many garbage containers there were so that, when they went back to measure, they wouldn't miss anything. They noted this information in their worksheet, as shown in Figure 6.2. They also noted that garbage was being generated in the kitchen and washrooms.

Garbage Waste Elimination Worksheet			
Activity or Area:	***Overall Building***		
Current State			
Identify	**Measure**		
Item	Material	Quantity	Hazardous Substances
Operations (6)			
Sales (8)			
Engineering (10)			
Accounting (6)			
Kitchen (2)			
Washrooms (4)			

Figure 6.2 Garbage waste elimination worksheet. Current state: overall building.

Step 2: Measure the Makeup of Garbage in Your Value Stream

It's time to get dirty! Once you know the areas where you are creating garbage, it becomes easier to focus on the composition of the garbage and how much of it there is. Once again, to start this step, go to the first activity of your value stream or the area that houses the activity you are focusing on. You have already identified the garbage cans, bins, and bags where the garbage for this area is being deposited, and you now want to look at what is in those garbage cans. Is it packaging garbage? What kind of packaging? Paper, plastic, metal? Is it leftover materials that can no longer be used? What are the materials made of? The key here is to list each of the individual items or materials that make up this garbage. For instance, if there is paper, plastic, metal, and wood, note each item separately. This will allow you to focus on each of those items, making it easier to minimize and eliminate them in later steps.

Once you know what items make up the garbage, you can measure them by either weighing each of the items or materials or counting them at the end of each day, week, or month. For example, if there is wood, how much wood or how many pieces of wood? How much plastic? And so on. Measuring will highlight the items and materials that contribute the greatest amounts and are costing you the most money. If you are having trouble with this task or don't want to do this yourself, ask your waste removal company to help you; many waste removal companies now offer garbage audits, which is essentially what this step is. Use this same process to measure the makeup and amount of overall building garbage. Note that you may want to do it for a couple of days, weeks, or months in a row to get an average amount of garbage, so that the data is not biased or skewed because of a high or low production period. Once you know what materials make up the garbage and their associated quantities, note it in your worksheet, as shown in Figure 6.3.

Having identified that garbage was being created in the forming activity of their value stream, Greanco green team members looked inside each garbage container to see what materials it contained. Opening up Can 1, they saw plastic shrink-wrap and plastic banding. They separated the items into two piles and weighed each pile. The shrink-wrap weighed 14 pounds, and the banding weighed 6 pounds. In Can 2, there were small metal pieces.

Garbage Waste Elimination Worksheet			
Activity or Area:	***Value Stream Activity 1: Forming***		
Current State			
Identify	**Measure**		
Item	Material	Quantity	Hazardous Substances
Garbage can 1	*Plastic wrap, banding*	*14 lbs, 6 lbs*	
Garbage can 2	*Metal pieces*	*29 lbs*	
Garbage bin 1	*Wood pallets*	*23 lbs*	
Oil container	*Hydraulic oil*	*86 lbs*	

Figure 6.3 Garbage waste elimination worksheet. Current state: value stream activity 1.

They poured all the pieces onto the scale and weighed them; it amounted to 69 pounds. In the garbage bin were wooden pallets. Instead of weighing the pallets, they counted them and came up with 23. For the oil, they first weighed a similar empty container, and then put the container with the oil in it on the scale and subtracted the empty container weight from the total weight to get the weight of the oil. They noted the materials that made up the garbage on their worksheet, as shown in Figure 6.3.

The Greanco green team then did the same for the overall building to determine the material makeup and quantity of garbage. They looked in each garbage can within each area and listed the garbage found in them, as shown in Figure 6.4. For the most part, it was paper and packaging materials, except for the kitchen, where there was plastic cutlery, plates, and organic waste.

Step 3: Measure the Hazardous Substances in Your Garbage

This step is important because it helps you with later steps and allows you to associate environmental impact with the garbage you create. This in turn allows you to focus on minimizing and eliminating the items that will bring you the greatest benefit. Hazardous substances are the most expensive to dispose of, as they require special handling, transportation, final disposal,

Garbage Waste Elimination Worksheet			
Activity or Area:	***Overall Building***		
Current State			
Identify	**Measure**		
Item	Material	Quantity	Hazardous Substances
Operations (6)	*Copy paper, card-board folders*	*2 lbs, 4 lbs*	
Sales (8)	*Copy paper*	*3.5 lbs*	
Engineering (10)	*Copy paper, plastic packaging*	*5 lbs, 8 lbs*	
Accounting (6)	*Copy paper*	*3 lbs*	
Kitchen (2)	*Organic waste, plastic cutlery, packaging*	*17 lbs, 6 lbs, 12 lbs*	
Washrooms (4)	*Paper towels, empty cleaning supplies*	*8 lbs, 4 lbs*	

Figure 6.4 Garbage waste elimination worksheet. Current state: overall building.

and so on, which cost more money than disposing of nonhazardous substances.

To determine whether garbage is hazardous, look at each material or item you identified in the last step. If it consists of leftover materials that went into the process, such as an offcut or defect, you will have already assessed it and given it a color rating during the Assess step in Chapter 5. If you did not do an assessment, or if the items are not a by-product of material input (packaging, for example), you will need to perform the Assess step in Chapter 5 to determine whether it is a hazardous substance. If the material is legislated by municipal or federal law, then you should already know whether the item is considered a hazardous substance. If you don't know yourself, ask the materials/purchasing/plant manager; they should be aware of what materials are considered hazardous. Another way is to look at your material safety data sheets (MSDS). If the item in the garbage has an MSDS, chances are it is (or has the potential to be) hazardous. If no one knows whether a material is regulated, ask the supplier or manufacturer of the product; they are required to know whether it is considered hazardous. Assess each of the items or materials you identified as making up the garbage, and note the hazardous substances on your worksheet, as shown in Figure 6.5.

Garbage Waste Elimination Worksheet			
Activity or Area:	***Value Stream Activity 1: Forming***		
Current State			
Identify	**Measure**		
Item	Material	Quantity	Hazardous Substances
Garbage can 1	*Plastic wrap, banding*	*14 lbs, 6 lbs*	*N/A*
Garbage can 2	*Metal pieces*	*29 lbs*	*N/A*
Garbage bin 1	*Wood pallets*	*23 lbs*	*N/A*
Oil container	*Hydraulic oil*	*86 lbs*	*Hazardous material (86 lbs)*

Figure 6.5 Garbage waste elimination worksheet. Current state: value stream activity 1.

The Greanco green team members compared all the garbage found in the value stream to the color criteria in Chapter 5, and against local and federal legislation, and found out that the hydraulic oil was considered hazardous by both sets of criteria. They noted this on their worksheet, as shown in Figure 6.5.

When the team members looked at the garbage being generated by the overall building and compared it to the two sets of criteria, they did not find any hazardous substances, so they noted that on their worksheet, as shown in Figure 6.6.

Step 4: Minimize the Creation of Garbage

Now we can start to take action. Minimizing the amount of garbage you produce will have an impact in a couple of different ways. First, it reduces the amount of garbage that you pay for, and then pay again to have taken away, bringing immediate economic savings. Second, it aids you in performing the next steps because you have less garbage to deal with. The environmental benefits are massive, because you will not only minimize the amount of garbage going to landfills, but also minimize the embedded green wastes involved with making and disposing of the item you are throwing away.

Garbage Waste Elimination Worksheet			
Activity or Area:	***Overall Building***		
Current State			
Identify	**Measure**		
Item	Material	Quantity	Hazardous Substances
Operations (6)	*Copy paper, cardboard folders*	*2 lbs, 4 lbs*	*N/A*
Sales (8)	*Copy paper*	*3.5 lbs*	*N/A*
Engineering (10)	*Copy paper, plastic packaging*	*5 lbs, 8 lbs*	*N/A*
Accounting (6)	*Copy paper*	*3 lbs*	*N/A*
Kitchen (2)	*Organic waste, plastic cutlery, packaging*	*17 lbs, 6 lbs, 12 lbs*	*N/A*
Washrooms (4)	*Paper towels, empty cleaning supplies*	*8 lbs, 4 lbs*	*N/A*

Figure 6.6 Garbage waste elimination worksheet. Current state: overall building.

Focus first on minimizing the hazardous substances, as they tend to bring the greatest economic and environmental benefits. In order to minimize the costs and environmental impacts associated with the amount of garbage you create (the "waste" of garbage), we will use a two-stage approach: minimize the physical amount of garbage being created, and then recycle or compost whatever you can. The order is not set in stone; do it in whatever order brings you the greatest economic benefits, because usually what brings the greatest economic benefits also has the greatest environmental benefits.

To minimize the physical amount of garbage you create, focus on one item or material at a time and ask this question for each: Where did this garbage come from? Typically, it will come from three general areas:

- Packaging
- Leftover or unused materials of a process
- Used-up items such as tools, jigs, oils, and solvents

To find out where your garbage is coming from, observe the activity and see it for yourself, or ask the employees in the area. For example, if the item is plastic, ask, "Did that plastic come from packaging?" If so, note that it came from packaging of a material. Is it leftover after a part is stamped or extruded? If this is the case, note that the garbage came from the stamping

Garbage Waste Elimination Worksheet				
Activity or Area:	***Value Stream Activity 1: Forming***			
Future State				
Minimize				
Item	Source of Garbage	Quantity	Recycle/ Compost	Savings
Garbage can 1	*Packaging*			
Garbage can 2	*Offcuts from forming*			
Garbage bin 1	*Packaging*			
Oil container	*Forming machine*			

Figure 6.7 Garbage waste elimination worksheet. Future state: value stream activity 1.

process. Once you have this information, note it in the Future State section of the Garbage Waste Elimination Worksheet, as shown in Figure 6.7.

To find out where garbage was coming from, the Greanco green team members stood and watched as the workers in the forming area performed their daily tasks. They saw that the plastic wrap and banding from Can 1 was the packaging material used to cover and strap the sheets of steel to the skids for transport from the supplier. The wood pallets were part of the packaging from the supplier as well. The metal pieces were the offcuts left over after the sheets of steel had gone through the forming machine. They could not see where the hydraulic oil was coming from, but after asking one of the operators in the area, they found that the forming machine had a hydraulic press in it and that the oil was changed on a monthly basis as part of the preventive maintenance schedule. They noted the sources of the garbage on their worksheet, as shown in Figure 6.7.

Determining the sources of garbage for the overall building took a little more time, but the team was able to do it easily enough. When they could not observe it themselves, they asked the people who used the garbage cans or the people in the area and easily determined the source of the garbage. They noted each source on their worksheet, as shown in Figure 6.8.

Garbage Waste Elimination Worksheet				
Activity or Area:	***Overall Building***			
Future State				
Minimize				
Item	Source of Garbage	Quantity	Recycle/ Compost	Savings
Operations (6)	*Reports, order folders*			
Sales (8)	*Mockups*			
Engineering (10)	*Design drawings, samples*			
Accounting (6)	*Trial balances, etc.*			
Kitchen (2)	*Lunches, cutlery, etc.*			
Washrooms (4)	*Handwashing, cleaners*			

Figure 6.8 Garbage waste elimination worksheet. Future state: overall building.

At this point, you now know where garbage exists in the value stream activity and in the overall building; you know the type and amount of each specific item or material in the garbage; and you know where it came from. Now let's focus on actually minimizing the physical amount. If the garbage comes from packaging, see whether you actually need that amount of packaging or any packaging at all. If you can use less packaging or no packaging, contact your suppliers and work with them so that your materials are packaged according to what *you* require. If the garbage is a result of leftover material such as an offcut, work with your designers or engineers to figure out a way to phase out leftover materials.

For example, if you are stamping a piece of metal and leaving little offcut pieces, could you rearrange or reposition the stamping to use all the material and leave no garbage? Could you use a bigger, smaller, or differently shaped sheet or different shape to avoid offcuts or leftover materials? Probe employees in the area, who live and breathe the process every day, for ideas they may have. If the garbage is a result of used-up materials that are no longer any good, you can do a couple of things. First, see whether you can make them last longer so they are not thrown out as often. Or, see whether you can clean the item or refurbish it so that it does not have to be thrown away or thrown away as often. A great example of this is hydraulic oils. This oil is used for a certain period of time, after which it gets dirty, is thrown away, and new oil is purchased. This oil can be cleaned and reused over and over again for much less money than throwing it away and buying new oil. When

Garbage Waste Elimination Worksheet				
Activity or Area:	***Value Stream Activity 1: Forming***			
Future State				
Minimize				
Item	Source of Garbage	Quantity	Recycle/ Compost	Savings
Garbage can 1	*Packaging*	*No wrap needed*		
Garbage can 2	*Offcuts from forming*			
Garbage bin 1	*Packaging*	*Send back pallets*		
Oil container	*Forming machine*			

Figure 6.9 Garbage waste elimination worksheet. Future state: value stream activity 1.

you have come up with a solution for minimizing the garbage, note it on your worksheet, as shown in Figure 6.9.

The Greanco green team members decided to pick the low-hanging fruit and focus on minimizing the physical amount of packaging garbage in the forming area. They wondered why the steel was wrapped in shrink-wrap. Wrapping was not required by any internal policies, so they called the supplier. The supplier told them that they wrapped all products going out because some people stored material outside, and if it wasn't wrapped it could rust. However, because Greanco stores all material inside, they do not need any shrink-wrap. The team also asked if they could reuse the wooden pallets if they were sent back to the supplier. The supplier agreed to both requests, and the team totally eliminated these two sources of garbage. They also negotiated a small rebate based on the number of shipments per month, because they were saving the supplier the cost of the shrink-wrap and new skids for every shipment. The team noted these actions on their worksheet, as shown in Figure 6.9.

To minimize the garbage created by the overall building, use the same approach. Garbage on a building level will typically come from the same three areas as the garbage in your value stream: packaging, leftover or

Garbage Waste Elimination Worksheet				
Activity or Area:	***Overall Building***			
Future State				
Minimize				
Item	Source of Garbage	Quantity	Recycle/ Compost	Savings
Operations (6)	*Reports, order folders*			
Sales (8)	*Mockups*			
Engineering (10)	*Design drawings, samples*			
Accounting (6)	*Trial balances, etc.*			
Kitchen (2)	*Lunches, cutlery, etc.*			
Washrooms (4)	*Handwashing, cleaners*	*Electric, cloth RFQ*		

Figure 6.10 Garbage waste elimination worksheet. Future state: overall building.

unused materials, and used-up items. Once you have gone through the process as you did for your value stream activities and identified a solution for the overall building garbage, note it on your worksheet, as shown in Figure 6.10.

In reviewing the garbage created by the overall building, the Greanco green team members looked at paper and the plastic cutlery and realized that efforts from dealing with those on the front end (in Chapter 5) would drastically reduce the amount of this garbage, so there was no need to focus on it right now. Instead, they focused on reducing the large quantity of paper towels being thrown away. They came up with two solutions: electric hand dryers and cloth towels; they are now waiting on quotes from vendors to see which, if any, makes the most financial sense. They noted this on their worksheet, as shown in Figure 6.10.

Now that you have minimized the amount of garbage you are creating, try to recycle or biodegrade whatever is left over, both in your value stream and in the overall building. There have been major advancements in the ability to recycle different materials over the past few years, and you may be surprised at what can actually be recycled. Contact your waste removal company and ask them to advise you on what materials in your garbage can be recycled. Also, explore different waste removal companies that have the capacity to accept certain items based on their relationships with end-user companies. It is important to recycle as much as you can; it will not only save the environment, but also save you money because recycling bins are cheaper to take away than garbage bins. There are buyers for recycled materials, so the waste removal company passes part of that profit on to you in the form of lower pick-up costs. Some companies even offer large rebates if you have enough recycled material because the recycled materials provide a revenue stream for them. Next, see whether you can biodegrade any of the garbage you are creating. To do this, assess whether each material can be composted. If you don't know yourself, ask a local environmental organization or do a Web search for "can [material x] be composted?" Also, contact your waste removal company, as they may be able to help you.

Once you have figured out what materials you can recycle, you need to recycle them. Your waste removal company can also usually help you with this, or you can set up your own recycling program. To create your own recycling program, set up a bin for fiber-based recyclables, such as paper; one for mixed materials that can be recycled, such as plastics and metals; one for compostable materials; and one for garbage that cannot be recycled or composted. Color-code each bin, such as blue for mixed, gray for fiber, green for compost, and black for garbage or do it according to your local recycling requirements. There is no need to purchase fancy bins: Use what you have in the form of boxes, pails, drums, and so on, and then put a color-coded label on each, with examples of items that go into them. For each area that is producing garbage, you will need a set of these bins.

Next, determine where those bins will be emptied. If they go straight to the curb, make sure that whoever is picking them up knows what is in the bin. If you are emptying the smaller bins into larger containers, make sure you are emptying them into the proper containers so that recycled materials are not being thrown away and garbage is not getting mixed with recyclables. This may involve sitting down with your cleaners or janitors to educate them on where they should be emptying those bins.

Next, you need to educate the employees in your organization about the program. Review why you are doing this, the color-coding system, and what goes into each bin. This will go a long way toward making the program successful.

When it comes to composting, you may choose to do it yourself on-site, instead of sending it away. If you have enough area on your property, you

Garbage Waste Elimination Worksheet				
Activity or Area:	***Value Stream Activity 1: Forming***			
Future State				
Minimize				
Item	Source of Garbage	Quantity	Recycle/ Compost	Savings
Garbage can 1	*Packaging*	*No wrap needed*	*Recycle plastic banding*	*14 lbs*
Garbage can 2	*Offcuts from forming*		*Recycle metal pieces*	*29 lbs*
Garbage bin 1	*Packaging*	*Send back pallets*		*23 lbs*
Oil container	*Forming machine*			

Figure 6.11 Garbage waste elimination worksheet. Future state: value stream activity 1.

could set up a compost heap, and then use the rich compost soil for your office plants and landscaping. Another popular way to compost is by vermicomposting. In vermicomposting, a plastic bin of soil contains red worms that eat the compostable material and turn it into rich, fertile soil. This can be done right in your office; it produces no smell, attracts no bugs, and is a great way to make visible to people the lengths you are going to be green. It helps to create a green culture and gets people talking about other green activities; it becomes almost like the green version of the water-cooler discussion.

Once you have determined what can be recycled or composted, note it on your worksheet, as shown in Figure 6.11.

When it came to recycling or composting at Greanco, none of the members of the green team were very knowledgeable. They decided to call in a local not-for-profit environmental organization. This organization not only told them what they were able to recycle or compost, but also helped them set up a recycling and composting program. The team members noted the items that could be recycled or composted on their worksheets, as shown in Figure 6.11 for the value stream activities and Figure 6.12 for the overall building. As you can see, most of the garbage items, both inside and outside their value stream, can be recycled or composted, and they are now diverting a large portion of their waste from

Garbage Waste Elimination Worksheet				
Activity or Area:	**Overall Building**			
Future State				
Minimize				
Item	Source of Garbage	Quantity	Recycle/ Compost	Savings
Operations (6)	*Reports, order folders*		*Recycle paper*	*2 lbs, 4 lbs*
Sales (8)	*Mockups*		*Recycle paper*	*3.5 lbs*
Engineering (10)	*Design drawings, samples*		*Recycle paper, plastic*	*5 lbs, 8 lbs*
Accounting (6)	*Trial balances, etc.*		*Recycle paper*	*3 lbs*
Kitchen (2)	*Lunches, cutlery, etc.*		*Compost organics*	*17 lbs, 6 lbs, 12 lbs*
Washrooms (4)	*Handwashing, cleaners*	*Electric, cloth RFQ*	*Recycle paper*	*8 lbs, 4 lbs*

Figure 6.12 Garbage waste elimination worksheet. Future state: overall building.

the landfill. The "Savings" column shows the weight of materials being "saved" from going to the landfill, in other words, the amount of materials being recycled.

An important note is that when you recycle or biodegrade something, it is referred to as "diverting it from landfill." If you divert 95 percent or more of your garbage from landfill, you are considered to be a zero-waste facility. However, this terminology is misleading, as you are still producing garbage; it is just not going to the landfill right away. The "right away" part is key, because eventually most things that are "recycled" end up in the landfill at some point. They are actually "downcycled," meaning that not 100 percent is recycled. For example, when you recycle a plastic bottle or metal, only a percentage of that material is reused in another product. This means that, on the first recycling pass, 20 percent of the original 100 percent goes to landfill and 80 percent is used in another product. On the second pass, another percentage smaller than the 20 percent goes to landfill. Over time, 100 percent of the original item does end up in the landfill; it just takes longer to get there. Real zero waste comes after you complete the last step in the garbage

CASE STUDY: BELL COMMUNICATIONS RESEARCH

The separation of recyclables from your other waste is only one way to lower expenses. Reducing the waste produced in the first place could generate even more cost savings. Minimizing trash not only lowers disposal expenses, but also reduces the amount spent on office and other vital supplies. Bell Communications Research, based in Livingston, Illinois, employs about 7,200 employees. The company estimates that it saves about $30,000 a year in purchasing costs by encouraging the use of reusable mugs over disposable cups. The company saves an additional $1,000 per year by reusing interoffice envelopes as many as 39 times. Remember, this is savings in purchasing costs alone, and the savings in disposal fees are in addition to this. Also, in addition to lowering costs, reducing waste through a company environmental program can boost both a company's image and employees' morale. An increasing number of studies show that consumers consider a firm's environmental record when making purchasing decisions.

waste elimination process; however, recycling is a good first step. It saves you money, and it is better for the environment to do it than not to do it.

Step 5: Move toward 100 Percent Reusable or Biodegradable Garbage

Being able to achieve the goal of 100 percent reusable or biodegradable garbage will allow you to realize even greater savings beyond minimization while you are on the road to elimination. By reusing the garbage you create, you are saving on numerous levels. You are saving the environment by reducing (a) the amount of garbage placed in a landfill and (b) the need to pull raw material out of the ground. On the economic side, you are saving by not having to pay to have the garbage removed, either because you are reusing it yourself or because someone who *can* use it is paying you for the privilege or taking it away for free. If you are reusing it, then you save again by not having to pay for virgin materials on the front end. Whatever way you look at it, there are tremendous benefits from carrying out this step. So, let's get started.

At this point, we have minimized the amount of garbage that is created, and we are recycling or composting what we can. But there is still some garbage not being recycled or composted and going to the landfill. This step focuses on making that garbage reusable/recyclable or compostable, or, in other words, increasing the diversion rate of garbage going to the landfill. To start, look at your worksheets (see Figures 6.11 and 6.12). What are the items left that are not being composted or recycled? Focus on one item at a time, and then focus on how to either recycle/reuse or compost it. There are a few ways to approach this. First, see whether you could reuse it anywhere: maybe in your own product, maybe in packaging your product, or maybe around the plant or office. If you can't use it internally, look externally to see whether someone else can use it.

A good place to start is with the material-exchange programs that exist in many cities or regions throughout North America. These programs exist to link people who are looking for used, unconventional items or materials with those who have these items to supply. You will be shocked at what people are looking for and what they are willing to pay for it. To find one of these programs in your area, do a Web search for "materials exchange program [your city]," or ask your local environmental or government organizations to help you find the closest program in your area.

If you cannot recycle/reuse or compost an item internally or externally, the last approach is to find out why not, and then change it so that it can be recycled/reused or composted. Is the item too small or large, or is it the wrong shape or color to be recycled or used in another application? Could you change any of these characteristics so that it could be reused? Does it have a toxic or special coating preventing it from being composted? Could you remove and dispose of the coating and recycle the bulk of the item/material? Ask this question: What changes would allow you to recycle/reuse or compost that garbage item? Then, it becomes a matter of whether it makes financial sense to do it. Remember that even if it doesn't make financial sense today, tomorrow it might. For this reason, note in the Green State section of the Garbage Waste Elimination Worksheet, as shown in Figure 6.13, any solutions you will implement to make garbage recyclable or compostable, including any ideas that are not feasible at the present time.

Completing this step, or getting 95 percent of the way there, allows you to proclaim yourself as a zero-waste facility. There are many out there, and in reality it is not that hard to achieve. As we know, that doesn't actually mean zero waste, but rather zero waste going to the landfill right away. I urge you to be honest and disclose what you mean by zero waste, and you

Garbage Waste Elimination Worksheet					
Activity or Area:	***Value Stream Activity 1: Forming***				
Green State					
Elimination					
	100% Biodegradable or Recyclable		Elimination		
Item	Solution	Savings	Solution	Solution Cost	Purchase and Disposal Cost
Garbage can 1					
Garbage can 2					
Garbage bin 1					
Oil container	*Clean and re-use oil*	*Material cost*			

Figure 6.13 Garbage waste elimination worksheet. Green state: value stream activity 1.

will be rewarded by your customers and employees for being honest and not "greenwashing" your efforts.

Because the Greanco green team was recycling or composting almost all of the garbage created by the overall building, the team members focused on the value stream activities. The hydraulic oil was deemed hazardous, so they focused on a way to deal with this. After doing a little Web research, they found out that they could clean this oil and reuse it over and over again. The company that provided this service had a mobile unit and could do it on-site. This approach also saved Greanco money that was being used to pay high disposal fees and purchase new oil every month—a win-win situation across the board. They noted this solution on their worksheet, as shown in Figure 6.13.

Step 6: Move toward the Total Elimination of Garbage

Now that you have minimized the amount of garbage you are creating, you have minimized the amount of work needed to totally eliminate it. Also, you are now reusing some of your garbage or returning some of it to the ground as a nutrient, providing you with the immediate savings of reduced disposal costs. Now you can focus on not having to pay for it in the first place by totally eliminating it.

Some organizations complete the previous step, and then stop there. The problem with this is that, although you are saving money (and, yes, maybe you have even started to create a revenue stream by selling what was once considered garbage and just thrown away), you still have a green waste and can still improve. If you are getting rid of garbage, even if you have found someone to buy it or you are composting it, you didn't need it in the first place. (There are exceptions, such as used-up parts, but we will focus on other garbage in this step.) For example, you have steel offcuts that you previously were throwing away or, perhaps, were even recycling, but now you have found someone who can use them as-is and is paying you for them. This is great—it's amazing, in fact—and it's much better than before, but how much are they paying you for them, and how much did you pay for them initially? Chances are, you are getting much less than you paid, so you are losing out. By totally eliminating garbage, you can eliminate this cost and reap huge savings.

To totally eliminate garbage, you can continue or step up your minimization efforts in hopes that you will eventually get to total elimination. As long as you are continually moving forward, this is an acceptable approach. On the other hand, you can take the next step and ask what it would take to completely eliminate garbage. There is an answer to this question. If you had no choice but to completely eliminate garbage, you could do it; anyone could, it just might cost money to do so. To answer this question, develop a solution internally by involving engineers or employees working in the area. It may require going to an outside third party who can help you find a solution. Check on what others in your industry or related areas have come up with. Whatever way you do it, remember that it needs to be done in a manner that does not produce garbage itself; otherwise the exercise is pointless. Once you have a solution to completely eliminate garbage, note it on your worksheet, as shown in Figure 6.14.

Because this was the Greanco green team's first shot at totally eliminating garbage, they wanted to start small, so they focused on the smallest item of garbage (in terms of amount or weight, not physical dimensions): the plastic banding. The question of how they could eliminate this garbage led to another question:

"Why do we need it, what purpose is it serving?" The answer was pretty simple, but to make sure, they checked with the supplier. Confirming their assumption, the supplier told them that the banding was needed to stop the sheets from sliding around during transport. With this in mind, they looked at other ways of accomplishing the same objective. One idea was to send the banding back to the supplier with their skids, so that it could be reused as well. Upon further investigation, they found out that banding would not last very long being reused, so they kept thinking. They came up with the idea that, if the sheets were in a box that was the same size as the sheets, they would not be able to slide at all. This could serve as a replacement to the plastic banding and maybe even do a better job. Also, they could turn the existing skids that were already on a reuse program into boxes by adding four pieces of wood. The skids could be used many times over, and there would be no more need for banding. Without looking at the costs of doing this, they noted this as a solution to how they could totally eliminate this waste, as shown in Figure 6.14.

Garbage Waste Elimination Worksheet					
Activity or Area:	***Value Stream Activity 1: Forming***				
Green State					
Elimination					
	100% Biodegradable or Recyclable		Elimination		
Item	Solution	Savings	Solution	Solution Cost	Purchase and Disposal Cost
Garbage can 1			*Turn pallet into box to eliminate plastic banding*		
Garbage can 2					
Garbage bin 1					
Oil container	*Clean and re-use oil*	*Material cost*			

Figure 6.14 Garbage waste elimination worksheet. Green state: value stream activity 1.

CASE STUDY: EPSON PORTLAND, INC.

Epson Portland, Inc., is a manufacturing company in Hillsboro, Oregon, that has received several awards for its waste diversion efforts. Currently, the company diverts 90 percent of its waste from landfill by recycling it, and the other 10 percent is sent to a waste-to-energy facility. The company's extensive recycling program includes the obvious recycling of cardboard boxes and other packaging, but also processes less obvious items, including liquid ink that is used for paint pigment, plastics, batteries, CDs, circuit-board scrap, printer cords and cables, along with wood pallets, to name a few. In 2000 alone, their recycling/reuse efforts resulted in 4.5 million pounds being diverted from landfill and translated into direct economic savings of more than $308,000, which came from money paid by recyclers or from the avoidance of disposal and tipping fees. This is in addition to the dramatic environmental benefits that go along with this effort.

Now that you have a solution for eliminating garbage, you need to determine whether it makes financial sense to implement it. First, look at how much you paid for the materials you are throwing away and how much it costs to have them taken away. To find out how much you paid for them, ask the purchasing or materials manager, or go to the supplier. To determine how much it costs for removal, estimate the percentage that item contributes to overall garbage and calculate that percentage of your disposal bill. If it is not part of the general garbage removal, i.e., if the garbage item you are focusing on has separate or added disposal costs, factor that into the cost. Add these two costs together and note the result on your worksheet, as shown in Figure 6.15.

The Greanco green team asked the supplier for a figure to represent how much they would be paying for plastic banding. The supplier said it was basically at cost and negligible, somewhere around 50 cents per skid. The disposal cost for the banding was so small they used the 50-cent purchase price. The team noted this on its worksheet, as shown in Figure 6.15.

Garbage Waste Elimination Worksheet					
Activity or Area:	***Value Stream Activity 1: Forming***				
Green State					
Elimination					
	100% Biodegradable or Recyclable		Elimination		
Item	Solution	Savings	Solution	Solution Cost	Purchase and Disposal Cost
Garbage can 1			*Turn pallet into box to eliminate plastic banding*		*$0.50 per skid*
Garbage can 2					
Garbage bin 1					
Oil container	*Clean and re-use oil*	*Material cost*			

Figure 6.15 Garbage waste elimination worksheet. Green state: value stream activity 1.

Now you need to determine the cost of implementing the solution for eliminating the garbage. You may have to get a quote from someone who could potentially supply the solution or, if you're doing it internally, come up with an estimate yourself. Note the cost on your worksheet, as shown in Figure 6.16.

The cost for the Greanco green team to make boxes out of the skids involved only a few minutes of labor to cut the wood and the cost of the wood itself. The cost of the wood was minimal, at $4 to $5 per skid. However, it was still a cost, and team members noted it on their worksheet, as shown in Figure 6.16.

Once you have determined the cost of implementing the solution and you know the cost of buying and disposing the garbage, do a simple payback calculation to determine how feasible it would be to completely eliminate garbage. You may be surprised. If it doesn't make sense right now, revisit the solution in six months or a year and do the calculation again; with constantly rising material prices, the solution may make sense not too far down

Garbage Waste Elimination Worksheet					
Activity or Area:	***Value Stream Activity 1: Forming***				
Green State					
Elimination					
	100% Biodegradable or Recyclable		Elimination		
Item	Solution	Savings	Solution	Solution Cost	Purchase and Disposal Cost
Garbage can 1			*Turn pallet into box to eliminate plastic banding*	*$5 per skid*	*$0.50 per skid*
Garbage can 2					
Garbage bin 1					
Oil container	*Clean and re-use oil*	*Material cost*			

Figure 6.16 Garbage waste elimination worksheet. Green state: value stream activity 1.

the road. You could also determine ahead of time what price the material would need to be for the solution to make sense.

> The Greanco green team members reviewed the numbers and realized that, after the new skid was reused ten times (it was capable of much more), it would pay back the $5 cost for the wood. The supplier agreed to add this to the shipment-based rebate program they were already on, whereby for each shipment Greanco would get an added 50-cent rebate.

Remember, even if it doesn't make sense now, keep working on minimizing garbage to move yourself closer to eliminating it. If you have completely eliminated garbage in your organization, that is truly zero waste and something to tell the world about.

Chapter 7

The Fifth Green Waste: Transportation

One of the greatest impacts we have had on the environment over the last few decades has come from transportation. Transportation includes the transport of humans from one place to another, as well as the transport of materials, supplies, and finished goods from one location to another. Whether it be driving a car to work, flying on a plane, or shipping products on a truck or boat, transportation has significant negative impacts on the environment. Looking at your value stream and your organization from the perspective of the environment, and reviewing transportation as one of the wastes, will allow you to see the effects and come up with a plan for eliminating the negative environmental impact and enjoying the economic benefits that come along with reducing transportation costs.

Unlike many of the other wastes you identify in your value stream or organization as a whole, it is a little harder to see the economic benefits that go hand in hand with the environmental benefits that come from eliminating the waste of transportation, but they are definitely there. Tax and other incentives for operating with an environmentally friendly fleet of vehicles add to the direct savings that result from using a hybrid or fuel-efficient vehicle. Also, there soon will be incentives in place for the purchase of carbon offsets to compensate for the negative environmental impacts caused by travel. In addition, by minimizing or eliminating travel and transportation, you won't have to pay for it anymore. How much do you spend each month on business trips or freight for shipping materials or finished goods? For most organizations, this is a staggering number, and by eliminating it or even minimizing it, you will have a huge impact not only on the environment, but also on the bottom line. With the soaring prices of fuel and no end in sight for these increases, most businesses could save thousands to millions of dollars by looking at transportation as a waste and working to minimize and eventually eliminate this waste.

In addition to the direct costs savings, transportation tends to be in the public eye, so minimizing and greening transportation will also be in the public eye, bringing with it the indirect benefits of going green, such as customer and employee retention and attraction. We will use a step-by-step process to tackle this waste as we have with the other green wastes. Because transportation waste stems from the fact that we travel in excess and in a manner that is harmful to the environment—resulting in excess costs, taxes, levies, and fees—the end goal in eliminating this waste is removing 100 percent of the negative environmental impacts caused by transportation, which is done by using ecofriendly modes of transportation. Again, we will look at intermediary steps you can take to minimize transportation waste in order to immediately realize environmental and economic savings.

The process for eliminating travel and transportation waste is:

Step 1: Identify the activities in your value stream and overall building requiring travel or transportation.
Step 2: Measure the mode of travel or transportation and the distances traveled or transported.
Step 3: Minimize the distances traveled or transported.
Step 4: Offset remaining negative environmental impacts of travel and transportation.
Step 5: Move toward 100 percent use of environmentally friendly modes of transportation and travel.

INTENDING TO USE ECOFRIENDLY MODES OF TRANSPORT

Having the intention to use ecofriendly modes of transport provides a framework for you to start coming up with ideas. It also gives meaning to the steps involved in eliminating this waste and provides the motivation needed to carry out all these steps, so that you reap the huge savings that are waiting.

Step 1: Identify Transportation within Your Value Stream and Overall Building

In this step, as with the other wastes, you will identify the activities that require transportation and not worry about the mode or distances. However, we will first break this waste into the two components of external transportation and internal transportation. External transportation is that which takes place outside of the organization's four walls, such as shipping and receiving items or business travel. Internal transportation consists of the travel required to move materials or products from one area or process to another. To identify external transportation, you will focus on the shipping/receiving activities of your value stream as well as the overall building activities. In identifying internal transportation, you will focus on the transport between the individual processes within your value stream and the overall building. We will start with identifying external transportation, as this will make up the majority of the overall transportation.

To identify external transportation associated with your value stream, approach it in one of two ways. First, try to identify all the things that are transported to you and that you transport to your customers. This includes raw materials that go into your product as well as supplies (machine parts, tools) required to transform your product into something of value, and then the finished goods you send to customers. You will need to go to the shipping/receiving area of your value stream and note the materials that have come in and the goods that are going out. To identify all the materials, you may have to look at a receiving log or talk to the shipper/receiver; you could also ask the purchasing or materials manager to help you list the items that have been transported to you. You may also want to look at your BOM, as this will list all materials going into your product. If you have too many items to list, you may choose to focus on the most common or high-volume items for the first pass. The second way to approach this is to get your incoming and outgoing freight bills and determine what was transported. Either way works. It is a matter of preference, but taking it from both sides will ensure the greatest accuracy in capturing all the things requiring transport in your value stream. Do not overdo it on the first pass; you are trying to identify those things that have been transported to you from a supplier and that you are transporting to your customer. If you miss a few things, you will pick them up as you go through the steps.

Transportation Waste Elimination Worksheet		
Activity or Area:	***Receiving (External Transportation)***	
Current State		
Identify	**Measure**	
Item	Mode	Distance
Steel sheets		
Hardware		
Plastic cups		
Paint		
Packaging		
Seat covers		
Sub assemblies		

Figure 7.1 Transportation waste elimination worksheet. Current state: receiving (external transportation).

Once you have identified the items that are being transported into your value stream from your suppliers and out of your value stream to your customers, note it in the Current State section of the Transportation Waste Elimination Worksheet, as shown in Figure 7.1.

The Greanco green team members went to the receiving area of the warehouse to see what they could find. There were a few items being unloaded from a truck, but they knew that this did not represent all of the items needed to make their products. They asked the shipper/receiver for help in identifying the regularly unloaded items. The shipper/receiver listed a few items. To ensure that they had identified as many items as possible, the Greanco green team also approached the materials manager to ask what items were most commonly being delivered. In doing this, they were able to capture a bulk of the items being transported to them. They noted these items on their worksheet, as shown in Figure 7.1.

Determining the external transportation required in the overall building used to support the value stream is a little more involved, but is approached in the same two ways. You can do it by looking at what materials are coming in—in this case, it is probably mostly office supplies—and noting them. You will also want to look at what is going out of the office via courier, mail, and so on, and note those things as well. Next, you will want to look at business travel, both incoming and outgoing. Finally, you will want to include the travel required for employees to get to work every day. Alternatively, you could pull out your travel expense bills, FedEx and UPS bills, postage bills, and so on and determine what the bill was for. Either way works; the point is to list all the things that require transportation. Once you have these items, list them on your worksheet, as shown in Figure 7.2.

For the external transportation required for the overall building, the Greanco green team first asked the receptionist—who was responsible for sending out mail, arranging courier pickups, and so on—the most common items she was sending and receiving. Most of the material coming in was office supplies or mail, and most of the items going out were either miscellaneous documents, such as contracts and letters, or accounting documents, such as invoices. The team members also knew that their sales people traveled, so they noted business travel as a source of external transportation. And, of course, employees had to get to work, so they noted employee commuting as an item, as shown on their worksheet in Figure 7.2.

Transportation Waste Elimination Worksheet		
Activity or Area:	**Overall Building (External Transportation)**	
	Current State	
Identify	**Measure**	
Item	Mode	Distance
Office supplies		
Mail		
Miscellaneous documents		
Accounting items		
Business travel (reps 1, 2, 3)		
Commuting		

Figure 7.2 Transportation waste elimination worksheet. Current state: overall building (external transportation).

To identify internal transportation, go to the area housing the value stream activity you are focusing on. Look for raw materials or supplies being added to the process, and for the finished or partly finished parts coming out of the process and going to the next activity or process. Obviously, these materials must have arrived from somewhere, so chances are you have already identified many of them under external transportation; however, in this step we are looking for the added internal transportation required once they arrive at your site. For example, if you receive items at your receiving dock, and then a few of these items, such as steel, plastic, and wood raw materials, are transported from the receiving dock to the first activity, that is internal transportation, and you need to identify it as such. Next, if a stamped part then traveled from stamping to the next process, that would also be internal transportation to be identified under Activity 2. Another way to approach this is to draw a simple map (spaghetti diagram) of each item coming into your facility, track it from the time it comes in the door to the time it leaves the door, see how far it travels, and allocate a portion of the distance to each value stream activity.

Once you have identified the items being transported internally, note them on your worksheet (this will require a separate worksheet for internal transportation between activities), as shown in Figure 7.3. You may want to do this for the overall building as well if you notice a significant amount of internal transportation taking place there.

Transportation Waste Elimination Worksheet		
Activity or Area:	***Value Stream Activity 1: Forming (Internal Transportation)***	
Current State		
Identify	**Measure**	
Item	Mode	Distance
Steel sheets		

Figure 7.3 Transportation waste elimination worksheet. Current state: value stream activity 1 (Internal transportation).

To identify internal travel, the Greanco green team stood at Activity 1 (forming) and watched what materials were being transported into the area. The only items team members could see coming in were the steel sheets going into the forming machine. To make sure this was all that was being transported into the area, they asked one of the workers. Nothing else was being transported to the activity, so they noted that single item on their worksheet, as shown in Figure 7.3.

Step 2: Measure the Mode and Distance of Transportation

Once you have identified the presence of transportation within your value stream activities and/or the overall building, determine the type or mode of travel, such as airplane, truck, boat, or train. Then, break it down further to determine the distance traveled by each mode. Start by identifying the modes of transportation and distance traveled for external transport of items into and out of the value stream. Then do the same for the overall building, and finish this step by measuring the internal transport required for both the value stream and overall building.

To start measuring the external transport of the value stream, look at each item on your worksheet, as shown in Figure 7.3, and identify how it was transported. Did it get flown in on a plane, and then travel by truck to your destination? Did it travel by truck all the way? Was it shipped on a boat and then put on a truck?

To find the answers, first ask the purchasing/materials manager or logistics coordinator. Second, contact the freight company or the supplier and ask them what mode(s) of transportation it took to get there. Once you know the mode of transportation for each item on your list, determine the distance it was transported. To get this information, ask the company that sends you the freight bill. If they are charging you freight, they have to be able to tell you the distance those items traveled. Or, if you know where it is coming from, use Internet maps to determine the distance. Once you have the mode of transportation and distance traveled for each item you have identified, note it on your worksheet, as shown in Figure 7.4.

Transportation Waste Elimination Worksheet		
Activity or Area:	***Receiving (External Transportation)***	
Current State		
Identify	**Measure**	
Item	Mode	Distance
Steel sheets	*Truck*	*85 miles*
Hardware	*Truck*	*15 miles*
Plastic cups	*Truck*	*35 miles*
Paint	*Truck*	*10 miles*
Packaging	*Truck*	*15 miles*
Seat covers	*Boat/truck*	*1000 miles, 120 miles*
Sub assemblies	*Truck*	*20 miles*

Figure 7.4 Transportation waste elimination worksheet. Current state: receiving (external transportation).

Greanco green team members knew that all items arrived at their building by truck. However, they did not know whether some items came from overseas, requiring travel on a boat, or if they came by train part way and then were loaded on the truck. To find out, they spoke to the purchasing manager. She told them that the seat covers were shipped by boat from China to the local port and then transferred to a truck, but all other items were shipped straight to them by truck. They noted this information on their worksheet, as shown in Figure 7.4. To determine the distances transported, they asked the purchasing manager for the address of the supplier for each item and plugged that into an Internet mapping site to get the distance. For the seat covers, they asked the freight company how far the item traveled by boat, and then checked the Internet for the distance from the port to their location.

To measure external transportation for the overall building supporting the value stream, use the same approach as for the value stream. Look at the worksheet in Figure 7.4 and, for each item you have identified for the overall building, determine the mode of transportation. For business travel, note if it is via plane, train, or automobile. For commuting, note if it is via car, public transit, or walking. One way to determine this information is to send a questionnaire asking employees how they get to work or how they typically travel for business trips. For courier travel, call the courier company and ask them.

If you can't determine the mode of transportation, or if it varies widely, make your best estimate. For instance, if you send most of your courier packages halfway across the country for next-day delivery, chances are that most of the transportation is via plane. For most transportation, however, you should be able to be fairly accurate in terms of the mode of transport. Once you have determined the mode of transport, determine the distance traveled. Oftentimes you can determine this the same way you determined the mode of transport. You may send out a questionnaire that asks both the mode and the distance, for instance. And again, if someone is charging you for transport (courier companies, for example), they have to tell you the distance it was transported. And again, another way to determine distances if you know the start and end points is to use an Internet mapping site to determine the rough distance. Either way, the point is to be able to associate a mode of transportation and distance traveled for each item that has been transported so that you can focus on those items that have the largest transportation requirements, as they will bring the greatest savings. Once you have this information, note it on your worksheet, as shown in Figure 7.5.

Transportation Waste Elimination Worksheet		
Activity or Area:	***Overall Building (External Transportation)***	
Current State		
Identify	**Measure**	
Item	Mode	Distance
Office supplies	*Truck*	*25 miles per trip*
Mail	*Truck*	*60 miles per trip*
Miscellaneous documents	*Truck, plane*	*280 miles per trip*
Accounting items	*Truck, plane*	*280 miles per trip*
Business travel (reps 1, 2, 3)	*Car, car, plane*	*3500 miles, 1100 miles, 600 miles (per month)*
Commuting	*Car*	*950 miles (per day)*

Figure 7.5 Transportation waste elimination worksheet. Current state: overall building (external transportation).

To measure the transportation required for the overall building in support of the value stream, Greanco green team members had to put in a little more effort. They each took an item and set out to determine the mode and distance traveled. The office supplies were easy because they were delivered on a truck from a big-box store, and they used the address of the store to get the distance. Mail was a little tougher; it was received from a number of different people but mostly from suppliers sending invoices. Instead of listing individual items, they used the average distance of the suppliers, and because all but one was local, they assumed transport by mail truck. For outgoing documents, most were going to customers, so they used the average distance of their customers. But because their customers were spread across a wide area, they knew some items were traveling by plane and some by truck, so they noted that. When it came to business travel, they talked to the people traveling. A majority of the travel was to see clients, so they asked those who traveled most the average distance they traveled per month and the mode they took (plane/car/train). Because the modes and distances varied greatly, they decided to list by sales rep to get a more accurate measurement. And, finally, to get the average distance that employees commuted to work, they sent an e-mail asking for this information and totaled the results. They noted the distances on their worksheet, as shown in Figure 7.5.

To measure the internal transportation of your value stream and overall building, use the same approach for external transportation by looking at mode and distance, but get the information in different ways. To start, pick an item that requires internal transport. How is that item being transported? For example, it may be by forklift or by crane. The best way to do this is to observe it yourself. Of course, you could always ask the shipper/receiver, operations, or plant manager as well. Once you know how the item is being transported, determine the distance. To do this, you will probably have to measure it yourself, so get out the measuring tape or other measuring device and physically measure the distance the item was transported. Because it is internal transport, the distances should not be very long. (Remember, if it leaves your four walls, it is external transport.) Once you have determined the distance transported, note it and the mode on your worksheet, as shown in Figure 7.6.

Transportation Waste Elimination Worksheet		
Activity or Area:	***Value Stream Activity 1: Forming (Internal Transportation)***	
Current State		
Identify	**Measure**	
Item	Mode	Distance
Steel sheets	*Propane forklift*	*800 feet per day*

Figure 7.6 Transportation waste elimination worksheet. Current state: value stream activity 1 (Internal transportation).

To determine the distance that the steel sheets were transported from the receiving area to the forming area, the Greanco green team members measured by counting how many ten-foot squares there were between the receiving area and the forming area. The total was 80 feet, and they were doing this trip about ten times per day, for a total of 800 feet per day. They also noted that it was being moved by a propane-fueled forklift, and recorded this on their worksheet, as shown in Figure 7.6.

Step 3: Minimize Transportation

To minimize transportation waste, start by focusing on external transportation, and then finish by looking at internal transportation. External transportation is where the bulk of transportation waste lies, and therefore it makes sense to focus your efforts on this area first in order to realize the largest savings. To start the minimization process, pick the largest source of external transportation according to the distances traveled. Depending on whether that source is within your value stream activities or the overall building, consult the appropriate strategy for minimization as discussed in the following sections.

Minimize Transportation in Value Stream Activities (Shipping and Receiving)

Typically, you will minimize transportation waste in your value stream by focusing on one of the following areas.

Source and Produce Locally

Wherever possible, source or produce locally to avoid the freight charges associated with longer distances of travel. To source locally, look for local suppliers of the item you are focusing on. It may seem cheaper to purchase from supplier A, but when you add freight charges, it may end up being more expensive. Start by looking to see where your competitors purchase from, or by doing a little research on suppliers of that item in your area. On the other end of your value stream, shipping items out, could you produce locally by having a satellite operation located close to a large customer or market? Would the reduced cost of your product due to reduced shipping costs make you more competitive and outweigh the costs associated with setting up a satellite location?

Use Transportation Demand Management

Whenever possible, you will want to closely manage the demand for transportation from transportation coming in or going out. To do this, look at consolidating shipments of materials being received and shipped out. For example, this can be done by arranging to have one shipment per week as opposed to two or three. This may conflict with your just-in-time policy or add to inventory carrying costs, but it then becomes a matter of determining what makes the most sense. Another way to manage the demand for transport is to do a milk run as opposed to having separate drop-offs or pickup of items. Doing things such as setting up routes to avoid left turns (to minimize idling time) and scheduling deliveries outside of rush-hour times will also help you manage the demand. If you are stuck on how you can reduce demand for transportation, there are a number of organizations popping up that specialize in this area. To find them, do a Web search for "transportation demand companies."

Use Other Modes of Transportation

Different modes of transportation have varying economic and environmental costs associated with them. Shipping by plane is more expensive than by truck, but shipping by truck is more expensive than by rail. Obviously, time for shipment varies as well, so look into having items shipped in and out by the cheapest mode of transportation that still meets your lead-time requirements. See what your freight provider or competing freight providers can offer to help reduce the environmental and economic costs of transport.

Avoid Rush Orders

Of course there is always the need to have items delivered on a rush basis, but this should be avoided whenever possible. The reason is that rush shipments cost much more than standard shipments because of the increased costs of getting it there quickly by using a dedicated truck or sending it by air. Would keeping a safety stock avoid rush shipments? If you always have to rush-ship items either in or out, you need to take a look at why that is happening because it's costing you money and having a larger impact on the environment.

Minimize Packaging

This refers to both the amount of packaging that comes with an item and the way in which it is packaged. Because packaging adds weight to the item being shipped and freight costs are tied strongly to the weight of the item being shipped, minimize packaging as much as possible. To reduce packaging, refer to the minimization step in the last chapter on garbage to help you cut down the amount of packaging that comes with the items being brought in. Another factor in freight charges is the amount of space occupied by the items being shipped. Because of this, you want to minimize the space these items occupy. Could more items be put in a box or on a skid by being arranged differently?

The best results typically come from a combination of the preceding tactics, but whatever you choose to do—minimizing the distance traveled, as well as the weight and size of your incoming and outgoing items—will reduce your costs and the impact on the environment as a result of your transportation. When you have come up with a strategy, note it in the Future State section on your Transportation Waste Elimination Worksheet, as shown in Figure 7.7.

Transportation Waste Elimination Worksheet			
Activity or Area:	***Receiving (External Transportation)***		
Future State			
Minimize			
Item	Distance		Offsetting
	Solution	Savings	
Steel sheets	*Consolidate shipments*	*170 miles/week*	
Hardware			
Plastic caps			
Paint			
Packaging			
Seat covers	*Threshold of 4% fuel increase*		
Sub assemblies			

Figure 7.7 Transportation waste elimination worksheet. Future state: receiving (external transportation).

The Greanco green team members focused on the seat covers coming from China. The question they asked was, "Why are we buying from China?" To get the answer, the team spoke with the purchasing manager, and she gave them the expected answer of "price." The green team urged the purchasing manager to look at the total landed cost of these items, and the purchasing manager replied that she had already done that initially. The green team then asked if she had done it recently. After another evaluation, the purchasing manager said that the price had gone up, but it was still cheaper; however, if it were to go up any further it would be more expensive. The team members noted the threshold on their worksheet, as shown in Figure 7.7. Picking the next largest item, they focused on the steel sheets. Because they had a long-term relationship with this supplier, they were opposed to switching, even though it was possible. They then turned to the possibility of consolidating shipments. Upon analysis, it was feasible to receive a larger shipment once per week instead of three times per week. Because the delivery truck was traveling 85 miles per trip, this amounted to a distance savings of 170 miles per week for two deliveries. Although this was not the ideal just-in-time delivery program, the green team looked at what made the most sense. After some analysis, they determined that consolidating shipments did not negatively affect the process or flow of production, and that the cost savings justified taking this action. They noted this initiative on their worksheet, as shown in Figure 7.7.

To minimize the transportation associated with the overall building, first look at the preceding ideas, but we will also introduce an additional set of action items that are more relevant to overall building transportation. Again, for best results, it will probably require a combination of action items.

Minimize Transportation in the Overall Building

In order to minimize transportation waste in the overall building, refer to the techniques discussed for the value stream, and build on those using the techniques in the following sections.

Use Technology

Wherever possible, use technology to reduce transportation. For example, instead of flying halfway across the country, can you have a Web meeting or do a video conference call? It costs a little money up front to set up the hardware and software to be able to do this, but it will pay itself back rapidly and produce residual savings month after month and year after year. Before you fly to your next business meeting, ask yourself if it is really necessary to go. Could it be done via the Web or some other way? Of course there will always be times when it is necessary to be there in person, but many times, it is not. Also, instead of sending something via courier, could you send it electronically? This includes invoices; not only is it cheaper and saves paper and postage, but it also saves the environment and the item gets to its destination faster. This means that, with an invoice, you get paid faster. Many times, sending a soft copy electronically is an acceptable substitute for sending the physical hard copy; do this wherever possible.

Consolidate

If you can, consolidate things into one shipment or trip instead of several. For example, do you need to send a repeat customer a bill after every order, or could your send them an invoice once a week? Could you e-mail soft copies regularly and send one bulk shipment of hard copies? Also, instead of having a courier pick up daily or multiple times a day, scheduling them to come less frequently will greatly cut down on transportation waste. If you are traveling from California to New York for a meeting, are there multiple meetings you could conduct with different customers/suppliers so that you don't have to go back in a month or two, even if it means extending the duration of your trip?

Carpool

For travel that stems from employees commuting to the office, a great way to cut down is to encourage carpooling between employees. A way to encourage this might be to offer the best parking spots to the carpooling cars.

Telecommute

Another way to minimize commuting is to offer your employees the chance to work from home or telecommute. Not only does this save on transportation, it saves on office space, increases morale of employees, and so on. Very popular these days for economic and productivity reasons, it also has great environmental benefits.

The preceding ideas will help you get started with minimizing external transportation. Remember that by shifting your thinking and focusing on ways to minimize transportation using these suggestions, you will surely find ways to do this. The result will be reduced costs and reduced impact on the environment, once again proving that doing a good thing for the environment almost always results in economic savings. Once you have identified some ways to minimize external travel and transportation, note them on your worksheet, as shown in Figure 7.8.

Choosing the largest contributor to overall building transportation, the Greanco green team focused on business travel. After talking with the reps who traveled the most, team members discovered that the trips were mostly scheduled visits to clients mixed with some trade shows and emergency calls. After discussing how they could cut down on the travel, they decided to set up Web meetings for every other scheduled visit. Their clients agreed, and actually applauded their efforts. This cut their travel nearly in half and gave them more time to focus on

Transportation Waste Elimination Worksheet			
Activity or Area:	***Overall Building (External Transportation)***		
Future State			
Minimize			
Item	Distance		Offsetting
	Solution	Savings	
Office supplies			
Mail			
Miscellaneous documents			
Accounting items			
Business travel (reps 1, 2, 3)	*Web meetings every other visit*	*1300 miles/mth*	
Commuting	*Telecommuting*	*105 miles/day*	

Figure 7.8 Transportation waste elimination worksheet. Future state: overall building (external transportation).

continuous improvement and development. The green team members also wanted to see what they could do for commuting travel, as this was a large item as well. They were able to convince senior management that telecommuting potentially has a number of benefits, and they decided to launch a pilot program with three employees working from home one day per week and see how it went. This alone would cut travel by 105 miles per day. The employees involved were thrilled, and other employees were also excited and hoped they would be able to participate as well. After the pilot program trial period was over, management would make a decision on whether and how to move forward with the program.

To minimize internal transportation, organize your value stream and overall building activities into a cellular arrangement. This technique is a common tool used in lean manufacturing, whereby you organize sequential processes or activities into a cell so that the next downstream activity is located directly beside the previous or upstream activity. This is done by physically rearranging the location of each activity so that it is beside the next downstream activity. It may be as simple as moving machines or work areas to be beside each other as opposed to halfway across the plant. Or it may involve breaking down a number of value stream activities into groups and having each of those groups arranged into a cell. Also, consider the storage of materials. Place the materials used in each activity as close as possible to that activity or cell. By doing this, you cut down or minimize the physical distance traveled between each internal process. When you have decided how you will arrange your activities and what you need to move where, note it on your worksheet, as shown in Figure 7.9. Please note that moving toward cellular manufacturing is more involved and requires more effort than simply moving machinery, especially in larger, more complex facilities. However, the benefits both from a lean and green perspective will, in most cases, justify the time and effort required.

The Greanco green team members looked at how they could minimize the distance the steel sheets were transported between the receiving area and the first activity of forming. First, they reviewed whether they could reduce the number of trips, but it was already minimal. Next, they looked at physically moving the forming station closer to receiving, but that would increase the distance from Activity 1 to Activity 2. They did notice, though, that they could move Activities 3 and 4 closer together and save on transport, so they will look at that in the future. Next, they looked at moving the receiving area closer to forming. Although this would work for this particular situation, it would increase the distance of materials going to other activities. Then they noticed that there were several doors where materials could be received, but they were only using one of the options; the door being used was actually the farthest door from the forming area. So they asked the operations manager whether they could receive the steel sheets only at the east door, because this would cut 20 feet of the distance. The operations manager agreed, and they cut 20 feet off every delivery of materials from receiving to forming—not a huge savings, but every bit counts. They noted these solutions on their worksheet, as shown in Figure 7.9.

Transportation Waste Elimination Worksheet			
Activity or Area:	***Value Stream Activity 1: Forming (Internal Transportation)***		
Future State			
Minimize			
Item	Distance		Offsetting
	Solution	Savings	
Steel sheets	*Receive at east door*	*200 ft per day*	

Figure 7.9 Transportation waste elimination worksheet. Future state: value stream activity 1 (Internal transportation).

Step 4: Offset Remaining Transportation

Only after you have exhausted the ways to minimize (a) transportation and (b) the lag time to the implementation of 100 percent environmentally friendly modes of travel should you look to offset. Offsetting is really a band-aid solution that doesn't address the root problem. Addressing the root cause, as we all know, is the only way to effectively eliminate a problem. As with a weed, the only way to get rid of it for good is to get to the root and deal with that; trimming it, or "offsetting," means you will always have to deal with it. Having said that, offsetting does have some benefits. The environmental benefits come from the reduction of CO_2 in the atmosphere. The economic benefits are a little harder to discern, but will typically stem from indirect benefits related to customer and employee perception of the good you are doing by offsetting. Although offsetting is not the solution, it is better than doing nothing, and it does show some commitment to the environment, as you have to pay upfront to do it. This commitment to a cause will translate into higher customer and employee loyalty, as they will feel good about being involved with an organization that is trying to do something good for the environment.

Offsetting of transportation is very similar to the other types of offsetting for energy use or emissions, as you will see in Chapter 8. You should use caution in choosing where to buy your offsets, and it is encouraged that you source the gold standard of carbon offsets for travel and transportation, as described in Chapter 3. For more specific guidance in

CASE STUDY: PILLER'S SAUSAGES AND DELICATESSENS

Piller's Sausages and Delicatessens is a family-owned food-service distribution company. As the demand for Piller's products increased, the company responded by adding larger fleets and new stops to the delivery schedule. Of course, back then there was little attention to the cost of fuel and the environmental impacts associated with burning it. As time went on, Piller's knew that it could do things more efficiently and not only save fuel, but save time and be more efficient overall. Because the delivery routes and sequences were rather complex and Piller's had no experience in efficiently managing this, the company turned to Descartes, a transportation demand management company, to help them. Descartes was able to help Piller's to sequence its deliveries, plan its routes, and even give the rest of the organization insight into the deliveries, saving the need to call and check with drivers, for example. The result was that within three months the company had increased the productivity of stops per hour by 12 percent and had cut labor, fuel, and maintenance costs by 8 percent per year. This translated into being able to pay back the cost of doing all this 10 times annually. Now that's a pretty good return on investment, not to mention the great environmental impact.

choosing the appropriate organization to purchase your offsets from, consult the "Consumers' Guide to Carbon Offsets," as described in the Offset step in Chapter 3. Before purchasing offsets, you need to choose what you are going to offset, so pick an item from one of your worksheets. To determine the amount of offsets you will need to purchase in order to neutralize the environmental impact from travel and transportation, you need to figure out the amount of greenhouse gases (GHG) or CO_2 that is being discharged as a result of the travel and transportation. This is a four-step process:

- **Determine the mode of transportation:** Obviously, different modes of transportation discharge different amounts of pollution. Luckily, you have already determined the modes of transportation in Step 2, and they are recorded on your worksheet, as shown back in Figure 7.4 for the value stream and Figure 7.5 for the overall building.

- **Determine the distance traveled:** Of course, the longer the distance traveled, the more the pollution. You already know how to measure or determine the distance traveled, as you have done this in Step 2. But don't just take the distance you came up with in Step 2, because if you have minimized travel and transportation, you will have a shorter distance and less cost to offset. Make sure the distance is the most current distance being traveled.
- **Determine the fuel type:** For driving modes of transportation, it is important to know the type of fuel being used, because there are different levels of pollution for different types of fuel. So you will need to determine the fuel type: diesel or unleaded, for example. To do this, ask the freight provider.
- **Determine the amount of CO_2 or CO_2e being discharged:** Because buying an offset results in a reduction or prevention of one ton of CO_2, you need to know how many tons of CO_2 are being produced as a result of the combination of the mode of transport, the distance transported, and the fuel used. This is very simple. Just go to any offset vendor's Web site (if you don't know any, do an Internet search for "carbon offset"). Most, if not all, of these vendors have carbon calculators that are easy to use. You insert the three pieces of information into the blanks, click calculate, and it gives you the tons CO_2 or CO_2e (carbon dioxide equivalents) discharged from that specific transportation profile, how many offsets are needed to neutralize the environmental impact, and the cost to purchase those offsets.

Because you already have most of this information, this step should take only a matter of minutes. Once you know how many offset tickets you will need and the cost associated with purchasing them, note it on your worksheet, as shown in Figure 7.10. Of course, you will need to purchase these offsets; letting people know that you have done so will reap you benefits. Just don't shoot yourself in the foot by greenwashing. If you use 100,000 km of transportation and you offset 1,000 km, do not put out a statement claiming that you offset transportation. Be honest and say you offset 1 percent of transportation. You will get more out of an honest statement in the long run. If you are offsetting 100 percent of transportation, brag about it, but let people know that you understand this is a temporary solution and you are working on minimizing and using friendlier modes of transportation.

Transportation Waste Elimination Worksheet			
Activity or Area:	***Overall Building (External Transportation)***		
Future State			
Minimize			
Item	Distance		Offsetting
	Solution	Savings	
Office supplies			
Mail			
Miscellaneous documents			
Accounting items			
Business travel (reps 1, 2, 3)	*Web meetings every other visit*	*1300 miles/mth*	*1,800 lbs CO_2, $98 annually*
Commuting	*Telecommuting*	*105 miles/day*	

Figure 7.10 Transportation waste elimination worksheet. Future state: overall building (external transportation).

For the external transportation from the overall building, the Greanco green team took the net distance traveled after savings from minimization and plugged the information into a carbon calculator. For Sales Rep 1, the monthly CO_2 created by traveling 2,200 miles (3,500 miles – 1,300 miles of savings from minimization) was 1,000 lb., which required only one offset costing less than $6. On an annual basis, it would cost less than $72 (typically, CO_2 offsets are around $5 to $6 per ton). For the other two sales reps, it was even less because they were visiting local clients most of the time. Because the monthly numbers were so small, they decided to use an annual figure. The green team members were actually surprised at how cheap it was to offset, especially after going through the minimization step. They noted the amount of CO_2 being produced from business travel and the cost to completely neutralize business travel in their worksheet, as shown in Figure 7.10.

Step 5: Move toward the Use of 100 Percent Environmentally Friendly Transportation

It is almost impossible to eliminate all transportation within any organization. But that doesn't mean that the transportation required has to have a harmful impact on the environment. We saw that by offsetting, you are able neutralize this harmful impact. However, although there are benefits to offsetting, it does cost money every year. This is why it is important to shift your thinking and seek environmentally friendly forms of transport so that you can eliminate the environmental impacts while receiving yearly savings, instead of incurring yearly costs. The trick is that, although there are some things you can do for very little or no cost to switch to ecofriendly modes of transport, many of the solutions do involve higher up-front costs (right now, anyway), which you will recoup one or more years down the road. However, once that cost is recouped, you enjoy the savings for many years. We will look at ways to do this for external transport of the value stream and overall building, and then finish off this chapter looking at how this can be done for internal transport.

For external transport of items into and out of the value stream, choose a freight company that offers ecofriendly modes of transport. Although in the past it was rare to see this, there are a number of ecofriendly freight services available now. These fleets use hybrid delivery vehicles or trucks equipped to run on biodiesel. Search out the companies in your area and explore the cost of changing over to an ecofriendly freight company; the cost, or the savings, may surprise you. There are a number of incentives in place to do this, so the cost after savings and rebates may actually be less. With the soaring costs at the gas pumps, the gap is closing, and these options will inevitably become cheaper than the non-ecofriendly solutions currently offered. You will also want to take into consideration what it means to your customers to have their items delivered in an environmentally friendly way; this in itself may be worth the extra cost, if there is one. Once you have determined a solution for moving an item over to ecofriendly transport, note it on your worksheet, as shown in Figure 7.11.

Because a number of items were delivered via suppliers' own fleets of trucks, and because the Greanco green team members did not want to disturb things too much on the first pass, they decided to look at one item (plastic caps) being delivered via a third-party freight company. They searched for a freight company

Transportation Waste Elimination Worksheet		
Activity or Area:	***Receiving (External Transportation)***	
Green State		
Eliminate		
Item	Eco-transportation	
	Solution	Savings
Steel sheets		
Hardware		
Plastic caps	*Biodiesel-operated trucks*	*Neutral*
Paint		
Packaging		
Seat covers		
Sub assemblies		

Figure 7.11 Transportation waste elimination worksheet. Green state: receiving (external transportation).

that used alternative-fuel trucks to make deliveries, and found one that was using biodiesel. They contacted the freight company and, at first quote, it was more expensive. After talking with the freight company, the green team was able to match the current price by committing to one quarter (three months) of shipments. They noted this on their worksheet, as shown in Figure 7.11.

For moving the external transportation requirements of the overall building to ecofriendly modes of transport, look at the ideas presented in this chapter for the items you are receiving and shipping (such as documents and supplies). We will also explore some additional ways to move other transportation requirements of the overall building to ecofriendly modes of transportation.

Hybrid/Fuel-Efficient Vehicles

For company cars and even employee commuting, hybrid or more fuel-efficient vehicles are a great way to save both money and the environment. The up-front cost of hybrids has come down in the past few years, and with the rising costs at the gas pumps, the paybacks are now rapid. If you add in the many government incentives, oftentimes it becomes a no-brainer decision to choose the hybrid over the standard model. This is particularly the case for sales people who do a lot of driving. For employees who are commuting, you may offer incentives to get them to make that choice as well.

Public Transit

Public transit is another good solution for employee commuting. If you are located on a public transit route, then encouraging employees to use public transit, or even offering incentives such as flextime, will have a huge impact on the environment. If you are not close to a public transit route, consider offering a shuttle service to and from the closest public transportation location. It will also show employees your commitment to the environment, and as you know, there are benefits that go along with that.

Alternative Fuels

If you have an in-house fleet of vehicles or delivery trucks, you may choose to equip these vehicles to use alternative fuels that are cheaper and have more stable pricing. Fuels such as biodiesel are cheaper and have much less impact on the environment. Although they are more costly up-front to install, the cost is quickly recouped, and the savings for years to come make it a smart decision in many cases. For more information on alternative fuels, see Appendix C.

Bicycles

The good old bicycle—what a great way to save the environment and money! Whether it is encouraging employees to bike to work by providing bike racks and changing rooms, or using bike couriers to send packages short distances in large congested cities, bicycles are a great alternative to motorized transport.

With the rapid increase in the price of fuel, you will find that one or a combination of these solutions is—today, right now—the smarter financial decision. The environmental benefits that come along with it are an added bonus. If you had not shifted your thinking to look at things from the perspective of the environment, and, thus, focus on transportation as a waste, you may have missed out on this opportunity for savings. Whatever you choose to do to use ecofriendly transportation, note it in the Green State section of your Transportation Waste Elimination Worksheet, as shown in Figure 7.12.

Transportation Waste Elimination Worksheet		
Activity or Area:	***Overall Building (External Transportation)***	
Green State		
Minimize		
Item	Eco-transportation	
	Solution	Savings
Office supplies		
Mail		
Miscellaneous documents		
Accounting items		
Business travel (reps 1, 2, 3)	*Hybrids for next company cars*	*1 year payback $3000/yr savings after*
Commuting	*Employee challenge*	*150 miles per day eco-travel*

Figure 7.12 Transportation waste elimination worksheet. Green state: overall building (external transportation).

When it came to eco-transportation, Greanco green team members wanted to continue focusing their efforts on business travel and employee commuting. For business travel, they asked senior management to consider purchasing hybrids the next time lease renewals for company vehicles came up. Management agreed to consider this as long as it made economic sense, so the team noted it as a future project and the approximate date when it would take place. For employee commuting, they issued a challenge to all employees. Whoever traveled to work the most per month by public transit or by walking/biking would win one extra day of vacation. The challenge was well received, and although there was only one winner, a number of employees participated, with dramatic results. In total, from all employees, it averaged about 150 miles per day of eco-travel versus driving their car in by themselves. The program was such a success—and the employees enjoyed it so much—that it is continuing each month and expanding to first, second, and third prizes. The green team noted their efforts and results on their worksheet, as shown in Figure 7.12.

To move internal transportation to an ecofriendly mode of transportation, grab some ideas from these action items, and build on them by using human-powered transport instead of energy- and fuel-based transport. You can approach this in a number of ways. First, replace the use of forklifts, for example, by using a manual pump truck powered by a worker to transport items. If items are too heavy to transport manually, look at breaking down the load into smaller pieces so that it can be moved by a worker. Another approach is to use wheeled conveyors, so that you can slide the item between activities or locations instead of using an electric or propane-fueled forklift. Many times this will actually reduce the transport time as well, because you don't have to drive the forklift to the item, pick it up, move it, and drive the forklift back to its parking spot. Once you have come up with a solution for using ecofriendly transport for internal transportation, note it on your worksheet, as shown in Figure 7.13.

This one was simple for Greanco green team members. Instead of using the propane forklift to move the skids of steel sheets from the receiving area, they could easily use a manual pump truck. It didn't take any longer, and the weight was acceptable to be transported via manual pump truck. They were using the forklift because that's what they had always done, and they had never thought about doing it differently because it had always worked fine. They noted this solution on their worksheet, as shown in Figure 7.13.

Transportation Waste Elimination Worksheet		
Activity or Area:	***Value Stream Activity 1: Forming (Internal Transportation)***	
Green State		
Eliminate		
Item	Eco-transportation	
	Solution	Savings
Steel sheets	*Manual pump truck*	*600 ft per day of propane lift truck*

Figure 7.13 Transportation waste elimination worksheet. Green state: value stream activity 1 (internal transportation).

CASE STUDY: METROPLEX IRELAND, LTD.

Metroplex Ireland, Ltd., is a freight company with a fleet of 11 trucks and 22 drivers. The owners operate 24/7, 364 days a year, primarily servicing more than 100 McDonald's restaurants. They knew that there were cheaper and more environmentally friendly ways to operate, so they started to look for ways to reduce the cost and environmental impact associated with the fuel used in their trucks. They chose a two-tiered approach in which they first looked at ways to minimize the use of fuel through setting benchmarks for fuel efficiency, and then trained drivers to achieve the benchmark. They also moved toward more ecofriendly transport and switched over to using biodiesel (used cooking oil). Metroplex now uses more than 2,000 liters of biodiesel every month. This results in a savings of nearly 60 tons of CO_2 and more than $110,000 per year. All this was achieved without modifying vehicles or fuel systems.

Addressing the waste of transportation is extremely important, as it is one of the largest overall contributors to environmental degradation. Improving in this area will no doubt bring you direct financial benefits, but it will bring an even greater positive impact to the environment. By eliminating this waste, you are making a commitment to go down the green road, and it goes a long way in proving to customers and employees your organization's commitment to going green. This will set in motion a huge wave of movement in the green direction that will bring with it green people, a green culture articulating more green ideas, and fuel for the green movement within your enterprise, leading to amazing results.

Chapter 8

The Sixth Green Waste: Emissions

In reviewing your value stream from the perspective of the environment and after assessing the first five wastes, you have identified many of the negative environmental impacts associated with your value stream and with the overall building that houses your operations. After assessing the waste of emissions, you will be one step away from having identified a majority if not all of your negative environmental impacts.

In scouting out the other wastes, you have indirectly identified the emissions being created. For example, in looking for energy waste, you may have identified an oven powered by electricity. The use of electricity on-site creates emissions from burning coal or other forms of energy generation back up the line at the power plant. These emissions would have been captured in your assessment of energy waste. However, you have not yet identified the direct emissions created at your location, such as inside that oven by baking or cooking items containing volatile organic compounds (VOCs) or other toxins that are then exhausted into the atmosphere. In exploring the waste of emissions, you will look at the direct creation of emissions on-site, which would not yet have been captured by looking at the other wastes. As mentioned, this may involve things like the operation of an oven, smelting of metals, welding, or on-site burning of fuels. In terms of the overall building, you will be looking at HVAC (heating, ventilation, and air conditioning) systems and other sources of emissions created directly on-site as part of the operation of the building.

You will also review emissions caused directly by your product or service. You may have already identified the emissions from your product or service as a result of the manufacturing or production process, but not yet looked at the emissions created directly by the product after it is produced. For example, if you were a car manufacturer, you would have captured the environmental impact as a result of the production process for making that car, but the fact that your product is something that releases large amounts of emissions when it is used by the customer has not been identified in any of the previously identified wastes. Identifying and eliminating the emissions associated with your value stream and organization as a whole—as well as the emissions discharged by your product or service once it is being used by the customer—offers an opportunity to make a dramatic, positive impact on both the environment and the bottom line.

We know that emissions contribute heavily to pollution, and we know the environmental impacts associated with pollution such as global warming. Therefore, eliminating emissions created directly in the production process—and in the use of the product after production—will have positive benefits for the environment. We also know that there is often a financial burden to releasing these emissions, in the form of levies, fines, or other monetary penalties. By eliminating emissions, you not only have a positive environmental impact, but you also eliminate the financial burden, bringing, for some organizations, a huge positive impact on the bottom line. Again, doing the right thing environmentally brings with it financial rewards.

We'll attack this waste in the same format as the others and use a step-by-step process along with green tools and techniques fueled by a shift in thinking to, first, minimize this waste, and then move toward eliminating it.

The process for eliminating emissions as a waste is:

INTENDING TO ELIMINATE EMISSIONS

Having the intention to eliminate emissions is a very simple step, but shifting your thinking to actual strategies enables you to start moving in the direction of realizing environmental and economic benefits. If you are focused only on minimizing emissions, you may do things differently than if you are thinking ahead to the total elimination of emissions. In your journeys, you may come across a solution that you can use to eliminate emissions, but if you don't have that intention, you may not pay close attention to the solution.

Step 1: Identify sources of emissions in your value stream activities and overall building
Step 2: Measure the type and amount of emissions in your value stream activities and overall building
Step 3: Identify the presence of emissions caused by the use of your product or service
Step 4: Measure the type and amount of emissions from the use of your product or service
Step 5: Minimize emissions using pollution-prevention techniques, tools, and technologies
Step 6: Offset the remaining emissions
Step 7: Move toward the total elimination of negative emissions

Step 1: Identify the Sources of Emissions to Produce Your Product or Service

What you want to identify in this stage is all emissions and effluents being created on-site and the items or processes creating those emissions. We will start by looking at the value stream activities. A good way to begin is to walk out to the first activity of your value stream and see whether anything is being vented or discharged outside or into the air of the building. (You may be surprised what you are breathing in.) Welding and brazing fumes are examples of emissions that may be discharged inside the building. Also, look for machines or tools that have a gas tank; if you are burning fuel on-site, you are creating emissions on-site. If you have trouble with this, ask workers in the area or the plant or operations manager if they know. You may also ask the EHS (environmental health and safety) rep or the environmental compliance officer, as they will be required to know what is being discharged into the atmosphere. Once you have identified that emissions are being created and what is creating them, note it in the Current State section of the Emissions Waste Elimination Worksheet, as shown in Figure 8.1.

The Greanco green team members walked to the forming area, and they noticed a pipe coming from the forming machine going straight to the roof. The worker operating the machine told them it was an exhaust pipe. As the steel sheet went through the machine, it was heated a little so that the sheets would bend more easily and not crack; during this process, smoke was created and vented outside. Upon further inspection of the area, they did not see any other signs of emissions, but they also confirmed with the workers in the area and the operations manager that this was the only source in that area. They noted the source of emissions on their worksheet, as shown in Figure 8.1.

To identify the emissions and effluents being created from the overall building, use the preceding approach and look for those things that are discharging emissions into the atmosphere or inside the building itself. It may be a little harder to take this approach for the overall building, though, so you can take a different approach as well. Walk outside, stand back, and take a look at the building and its surroundings. Is smoke or steam emanating from your building? Does a haze surround your building or area? Are there smokestacks sticking out of the roof? There probably are, so if you don't see any, look more carefully. Once you spot where emissions are exiting the building, or where there is potential for emissions to be exiting the building (smokestack), get more specific and pin down the activity causing the emissions.

To do this, ask the building manager or facilities manager—or even the plant or operations manager—what that smokestack is attached to or what is creating the smoke coming out of the building. An obvious answer would be the HVAC system. Remember: At this point you don't have to measure the exact amount or makeup of the emissions; you are concerned only with

Emissions Waste Elimination Worksheet		
Activity or Area:	***Value Stream Activity 1: Forming***	
Current State		
Identify	**Measure**	
Item	Type	Amount
Forming machine (steel)		

Figure 8.1 Emissions waste elimination worksheet. Current state: value stream activity 1.

identifying that emissions exist and the item that is creating them. Once you have identified what is creating the direct release of emissions from your overall building (emissions that are separate from your value stream), note it on your worksheet, as shown in Figure 8.2.

Turning to the emissions from the overall building, the Greanco green team members asked the facilities manager to help them identify the emissions being caused by activities outside the value stream. The first thing he pointed out was the heating, ventilation, and air conditioning (HVAC) system. The team then asked what kind of heating it was, and he told them it was natural gas. He also told them that the system was put in more than ten years ago, and there had been frequent temperature-related complaints over the past year. Asked if anything else was being vented out of the facility, he pointed out only the value stream activities. Just to make sure, the team walked around the outside perimeter of the building and looked for any more smokestacks. They noticed a couple that had not been accounted for, but upon further investigation they discovered them to be exhaust fans from the bathrooms. For now, they noted the HVAC as the only source of emissions for the overall building, as shown in Figure 8.2.

Emissions Waste Elimination Worksheet		
Activity or Area:	***Overall Building***	
Current State		
Identify	**Measure**	
Item	Type	Amount
HVAC system		

Figure 8.2 Emissions waste elimination worksheet. Current state: overall building.

Step 2: Measure the Type and Amount of Emissions Used to Produce Your Product or Service

Once you have identified that emissions are being created and what is creating them, measure the type and amount of emissions. Start by looking at the value stream first, and then the overall building. Choose one of the emission producers in your value stream. There are a couple of ways you can go about measuring the type and amount of emissions being discharged. The first way is the easiest. If you are discharging toxic emissions into the atmosphere, you are probably required to have a certificate allowing you to do so. In order to get this certificate, you probably had to determine and prove the type and amount of emissions. Pull the file supporting this certificate, and you will have most of the answers.

The second way to go about it is to look at the materials causing the emissions. For example, if you identified the oven as a producer of emissions, then identify what in that oven is causing emissions to be produced. If you turned on the oven with nothing in it, you would just be exhausting hot air, but something in the oven is exhausting toxins that are contained in that material. Refer to the material's MSDS (material safety data sheets), which will tell you the type and amount of VOCs (volatile organic compounds) released into the atmosphere as a result of baking it or even using it in its intended way. Ask your EHS rep to show you the proper MSDS sheet for the material in question. If you do not have an MSDS sheet, ask the supplier or manufacturer to provide it (they must do so, legally). If you are using an item that will result in the release of VOCs or other environmentally harmful emissions, it most likely has an MSDS sheet or similar piece of information. You may have already done a lot of this work in the Assess step for materials waste, as detailed in Chapter 5, so refer to the data acquired to help you here as well.

The final approach is to have the emissions tested by a qualified third-party tester. You may want to skip to this step right away because it is the most accurate and current answer to the type and amount of emissions. To find such a tester, do a Web search on "emissions testing in [your area]," which should yield a number of organizations. You can also consult your local pollution prevention association, also known as a P2 organization. They will be able to perform the tests or point you in the direction of someone who can. Once you know the type and amount of emissions being discharged, note it on your worksheet, as shown in Figure 8.3.

For the first activity, forming, the Greanco green team members had only one thing to focus on: the exhaust from the forming machine. They asked the plant manager if there were any certificates of air emissions for the forming machine, and he replied that they were not required to have one. They then asked if the amount of emissions had ever been tested, and this had not been done either. Having exhausted the easy route, they turned to their supplier to find out whether steel produced any emissions as a result of their manufacturing process. The supplier did not know either, but gave them a contact at the steel mill who might be able to help. They learned exactly what they needed to know from the steel mill contact; because the stevel mill made the product, they were required to have this information readily available. They discovered that there were three main types of emissions that could be discharged if the steel was heated: sulfur oxides, nitrous oxides, and some fine particulate matter. The mill was also able to provide a rough estimate of the amount of each of these emissions per ton of steel. They noted this on their worksheet, as shown in Figure 8.3.

To measure the type and amount of direct emissions created on-site as a result of the overall building operations, you can refer to the techniques presented for the value stream activities. However, if you want to avoid testing by a third party, or if the other techniques for the value stream

Emissions Waste Elimination Worksheet		
Activity or Area:	***Value Stream Activity 1: Forming***	
Current State		
Identify	**Measure**	
Item	Type	Amount
Forming machine (steel)	*NO_2, SO_2, fine particulate matter*	*100 lbs/ton; 130 lbs/ton; 40 lbs/ton*

Figure 8.3 Emissions waste elimination worksheet. Current state: value stream activity 1.

do not work, talk with the people who provided you with these systems, whether they be HVAC or other systems. They should be able to tell you the type and amount of emissions created as a result of using their system or product. Pay attention to the burning of natural gas, heating oils, or other fuels on-site, because although some of these fuels burn fairly cleanly, some do not, and sometimes the fuels that normally burn clean are contaminated. If they are low-grade or contaminated fuels, they may actually be releasing toxic emissions into the atmosphere or possibly even inside the building. To find out the type and amount of emissions being discharged from the use of these fuels, contact the utility or other fuel provider and ask them. If you have exhausted ways to determine the type and amount of emissions being created with no success, you may have to resort to third-party testing.

It is also important to note that sources of indoor air pollution can include things like paints, carpeting, office furniture, or electronic equipment. Because the emissions from these things are essentially "invisible" and the sources are rather difficult to identify, you will most likely have to resort to having an Indoor Air Quality Test performed. This will tell you if there are any toxins being discharged inside the building, and it can be completed for relatively low cost. Taking this last step will ensure that you have captured all the emissions, both as a result of the value stream and the overall building, and allow you to focus on minimizing the biggest offenders. Once you have the type and amount of emissions from the overall building, note it on your worksheet, as shown in Figure 8.4.

Emissions Waste Elimination Worksheet		
Activity or Area:	***Overall Building***	
Current State		
Identify	**Measure**	
Item	Type	Amount
HVAC system	*NO_2, SO_2, fine particulate, CFCs*	*42 lbs/month; 18 lbs/month; 10 lbs/month; trace*

Figure 8.4 Emissions waste elimination worksheet. Current state: overall building.

To measure the amount of emissions coming out of the building, the Greanco green team members decided to take the easy route and have it tested along with the indoor air quality, as they deemed it to be a health and safety issue as well as an environmental one. Upon reviewing the test results, the green team was surprised to see that there were actually some sulfur oxides, nitrous oxides, fine particulate matter, and trace amounts of CFCs (chlorofluorocarbons). They noted this on their worksheet, as shown in Figure 8.4.

Step 3: Identify the Presence of Emissions from the Use of Your Product or Service

Having captured the emissions associated with all the activities required to bring a product or service to life, you can now focus on identifying the emissions created once your product or service is being used by your customer. Identifying the presence of emissions caused by your products and services once they are out in the market does take some extra effort, but it is well worth it. The environmental impact of your product once it is in use is what sticks in the customer's mind. Your customers have to live with the impact they have on the environment every day, so it makes sense that this is what they are most concerned about. Identifying this waste and moving toward minimizing and eventually eliminating it brings with it large opportunities. It is no secret that if two products or services can achieve the same end results, and if one is more environmentally friendly than the other, consumers are more and more often consciously choosing the one with less environmental impact. So although working on this may take up-front investments, the payoff in terms of customer loyalty and attracting new customers is worth it.

To identify whether emissions are being created by one of your products or services, ask the design engineers, the product manager, or someone else inside your organization that may know. If you offer a product or service that is creating emissions when used by the customer, chances are that you are legally required, at least, to have had testing done. If not, ask the design engineers or product manager or do a quick search for any information on the same (or similar) product or service you are providing. Once you have

Emissions Waste Elimination Worksheet		
Activity or Area:	***Finished Product***	
Current State		
Identify	**Measure**	
Item	Type	Amount
Furniture (paint, glue)		

Figure 8.5 Emissions waste elimination worksheet. Current state: finished product.

identified the products or services that do create emissions when in use by the customer, note it on your worksheet, as shown in Figure 8.5.

Because the only product Greanco makes is metal furniture, the Greanco green team could not see how their product would create any emissions once in the hands of their customer. The green team did remember, though, that in the later processes of the value stream they were using some paints and glues that contained VOCs. Looking into this a little further, the team members realized that these substances actually do release fumes for weeks to months after manufacture. Although they did not know how much, they did note it on their worksheet, as shown in Figure 8.5.

Step 4: Measure the Type and Amount of Emissions from the Use of Your Product or Service

Now that you know which products or services are creating emissions once they are being used by the customer, focus on measuring the type and amount of those emissions. This can be approached in a couple of ways. Of course, you can have it tested by a third-party qualified tester. However, if your product is the same or very similar to existing products or services, first see whether this has already been done. If it has, use that data to give you

Emissions Waste Elimination Worksheet		
Activity or Area:	***Finished Product***	
Current State		
Identify	**Measure**	
Item	Type	Amount
Furniture (paint, glue)	*VOC*	*Trace*

Figure 8.6 Emissions waste elimination worksheet. Current state: finished product.

a good idea of the type and amount of emissions generated from the use of your product or service. This will save you the time and money associated with doing it yourself, but still allow you to move toward minimizing and eliminating these emissions. Once you have this data, insert it into your worksheet, as shown in Figure 8.6.

After talking to the supplier of the paint and glue, the Greanco green team members realized that the amount of fumes was barely traceable by the time the products were in the possession of the customer. However, there were still trace amounts of various volatile organic compounds (VOCs) similar to those found in older paints and glues, so they noted it on their worksheet, as shown in Figure 8.6, because even though small, these trace amounts were still being released into the atmosphere and being inhaled by those using or coming into close contact with the product.

Step 5: Minimize Emissions

Minimizing emissions can be approached in two basic ways. The best way is to minimize them at the source, but if that is not possible, aim to clean them before discharge into the atmosphere. First, we will look at approaches for dealing with the source and then at cleaning emissions, for both your value stream and the overall building. We will finish by looking at ways to minimize the emissions created by the end use of your product or service. Performing this step will bring with it a number of economic and environmental benefits. The environmental benefits come from the fact that less emissions means less pollution contributing to global warming and other negative impacts, as discussed in Appendix B. The economic benefits come from reduced fines, levies, taxes, and other monetary penalties that come with discharging emissions. Also, because emissions tend to be very visible to employees and customers, minimizing emissions shows your commitment to going green, leading to increased retention of customers and employees as well as attracting new customers and employees. It also limits your exposure to the many forms of environmental or carbon taxing programs that are on the horizon.

Before getting into the specifics of what you can do to start minimizing emissions, make sure that things are operating as intended. It is very possible that the amount of emissions from your operations or your products/services is a result of a system not operating properly or not installed properly or completely. The best way to determine this is to have commissioning done at the time the equipment is implemented or to perform a checkup to ensure that all equipment is running properly. (Commissioning is a process whereby, upon startup of a system, various tests are performed to ensure that the system is functioning as promised.) To get the commissioning done, ask the supplier of the system to do this for you or consult the local contractor or representative responsible for installation. There are also a number of independent commissioning agencies that can do this for you.

To minimize emissions at the source, start by looking at one activity or item you have identified in your value stream or in the overall building as a source of emissions. In going through the Measure step, chances are that you did identify the root cause of the emissions. This may have been the use of an epoxy or varnish; or it may have been the baking of a certain material, or something of that nature. Whatever it is, focus on the material or substance that is the root cause of the emissions. Make sure that you are using only the required amount of that item or material. Any excess use beyond what is actually required results not only in excess amounts of that material being used, leading to higher costs, but also in excess emissions. Check how much is being used in practice versus how much is required as per the design. If there are discrepancies, you will need to take the necessary steps to use only the amount required. You can then proceed to the next step and ask whether there is any way to use less than is currently required by the design. If the design calls for ten units of something, could you use nine or eight units and still get the same quality? Can the concentration of the material or substance be diluted and still provide the same effect? Whatever the case may be, use only the bare minimum amount of this item.

Next, see whether you can substitute the item in question with a more environmentally friendly alternative. Talk with the supplier and see whether they have any alternatives you could test. Over the last decade, there have been rapid advancements in environmentally friendly alternatives to traditional materials and ingredients. You may be surprised at what choices exist today. If you can't find an alternative, let suppliers know that you are looking for one and are willing to work with them to come up with a solution. You can also research what your competitors or others have done in similar situations. Unless you are doing something unique, chances are someone has already tackled the problem and come up with a solution. And, of course, you can always talk to your local nonprofit environmental agencies. Many pollution prevention (P2) organizations are waiting to help you come up with ways of minimizing your emissions. It is important to emphasize that environmentally friendly alternatives are continuously developing, and if no alternatives are available today, one may be available tomorrow, so continue to look or revisit it in a few months to see what new solutions are out in the market.

Next, examine the possibilities for altering, changing, or removing the activity or process itself. Could air-drying substitute for baking an epoxy? Could you shorten cure time, resulting in less emissions being discharged? Could you adjust temperatures so fewer volatiles are released? Can you do away with the step altogether? These are the types of questions you need to be asking. If you don't know the answers, consult your engineers or operations specialists. You could also turn to local P2 organizations for help or look at what others have done in similar situations.

Once you have minimized emissions at the source, look at filtering or cleaning the emissions that are left. This is done by installing scrubbers or other types of filters at, or just before, the point of discharge to cut down the amount of emissions that are released into the atmosphere. Because the type of scrubber or filter varies depending on the type of emissions, you will need to consult with specialists in this field to find the right solution for you. If you have trouble locating someone, turn to your local pollution prevention organization to help find the right one for your needs. Remember that taking the approach of filtering or cleaning emissions will, in most cases, be the most costly and have the fewest benefits both economically and environmentally; it should be done only after minimizing the source of emissions. Once you have developed a solution to minimize the amount of emissions being discharged into the atmosphere, by dealing with them either at the source or at the point of discharge, note it in the Future State section of the Emissions Waste Elimination Worksheet, as shown in Figure 8.7.

The Greanco green team was stumped on how to cut down the amount of emissions created as a result of heating the steel in the forming machine. Upon further investigation, team members realized that some of the emissions were coming from the grease on the sheets. An easy fix involved wiping the sheets down with a dry cloth after they were sprayed. This helped a little, but most of the emissions were coming from the steel itself and the coating on the steel. Getting the engineers involved, the team learned that heating was necessary to avoid the steel cracking as it was being bent and formed. There was no way around it at this point. The green team then turned to the last resort of attaching a scrubber to the discharge pipe. This was a viable solution that would cut out more than 80 percent of the emissions being discharged; however, it involved a large up-front cost. Because there was other lower-hanging fruit, the green team noted this as an option on their worksheet, as shown in Figure 8.7, but decided to focus on other projects.

When it came to the emissions of the overall building, the Greanco green team was confused as to where the emissions were coming from. Team members asked the maintenance crew to look at the air conditioning system to see whether the refrigerant was leaking. Sure enough, there was a small leak leading to trace amounts of CFCs escaping into the atmosphere. The system had been installed before the ban on CFCs came into effect. This was quickly repaired. Next, they looked at the other substances being emitted and noticed that they were the same substances being emitted from some of the value stream activities. They checked whether the piping used to exhaust the emissions from those activities had a leak. Again, this was the case. These leaks were repaired, and the emissions dropped dramatically for the building but increased for the individual value stream activities. The team added inspection of these systems to the preventive maintenance program. It noted the repairs on its worksheet, as shown in Figure 8.8.

Emissions Waste Elimination Worksheet				
Activity or Area:	***Finished Product***			
Future State				
Identify	**Measure**			
Item	Discharge		Savings	Offsetting
	Source	Scrubbing		
Forming machine (steel)	*Wipe off sheets of steel*	*Install scrubber*		

Figure 8.7 Emissions waste elimination worksheet. Future state: value stream activity 1.

Emissions Waste Elimination Worksheet				
Activity or Area:	***Overall Building***			
Future State				
Minimize				
Item	Discharge		Savings	Offsetting
	Source	Scrubbing		
HVAC system		*Required leaks*	*Virtually all emissions*	

Figure 8.8 Emissions waste elimination worksheet. Future state: overall building.

Now that we have looked at how to minimize the emissions being discharged by your value stream and the overall building, we can focus on emissions created as a result of the end use of your product or service. Of course, this in itself has a large area for discussion and research, so getting into specific detail is beyond the scope of this book. However, we can look at ways to get moving in the right direction.

If you have a product or service creating ongoing emissions when used by the customer, you need to get to the root cause of those emissions. If the emissions result from the use of materials you are putting into the product, you will address this in Chapter 5 by going through the steps for eliminating materials waste. However, chances are that the emissions come from burning some sort of fuel or material. In this case, look at ways to avoid

Emissions Waste Elimination Worksheet				
Activity or Area:	***Finished Product***			
Future State				
Minimize				
Item	Discharge		Savings	Offsetting
	Source	Scrubbing		
Furniture (paint, glue)	*Phase out paint and glue*			

Figure 8.9 Emissions waste elimination worksheet. Future state: finished product.

those materials and fuels by using an environmentally friendly alternative. A well-known example of this is vehicles that run on alternative fuels such as biodiesel. Could your product run using an alternative fuel? Could it run on batteries, or could you offer a manual version of your product? Could it run using dedicated solar or wind power? Another approach is to educate customers who use your product about how they can do their part by using the product or service in the most efficient way possible in order to minimize the emissions created. If you show your customer how to lessen the emissions they create, you not only help them lessen their impact on the environment, you also save them money through the use of less fuel or materials. A great example of this is tips for increasing fuel efficiency; can you do the same for your product? Whatever solution you have come up with, even if it seems radical today, it might not be tomorrow, so note it on your worksheet, as shown in Figure 8.9.

Even though the amounts were very small, Greanco green team members were still concerned about trace emissions from their product, more for the safety of customers than for the environment. Because the emissions were a result of the materials they were using, they decided to deal with this by phasing out the use of these materials in order to eliminate it at the source. They noted this as their solution on their worksheet, as shown in Figure 8.9.

CASE STUDY: GUILFORD MILLS

Guilford Mills is a manufacturer of knit and woven fabrics for a number of different end-use markets. One of its plants in North Carolina performed a number of activities within its value stream, such as yarn spinning, weaving, washing, printing, and heat setting. After identifying and measuring the emissions created by those value stream activities, it was able to determine that they were largely from the solvent-based chemical system in the heat-setting process and from the printing activity. In the heat-setting process, a gummy solvent containing VOCs was added to the edges of the fabric as it went through the process to keep the edges from curling during the heating process. With a new piece of environmental legislation coming down the pipe that set a cap on emissions and charged for exceeding that cap, Guilford Mills wanted to see whether they could avoid that. After testing some ways that it could minimize their emissions, it found that a water-based solvent was a suitable alternative and required very little retooling. Through its minimization efforts, it was able to cut emissions from 246.8 tons per year to 93.7 tons per year. This put Guilford Mills below the threshold cap set by the new legislation, saving it $6,000 per year. The cost of retooling and capital investment was paid back in six months.

Once you start reducing emissions, make sure you take the proper action to have your levies or fines reduced and to inform your customers, employees, and shareholders of the good things you have done to reduce their impact on the environment. Once again, you'll quickly start to see that going green really does have multiple benefits associated with it. For some organizations, this can result in huge dollar savings that most of us would never imagine. By tackling the other wastes and moving toward the elimination of them, you will unavoidably also have an impact in minimizing and eliminating this specific waste of emissions.

Step 6: Offset Remaining Emissions

Just as you can offset the negative environmental impact of energy and transportation, you can offset the negative environmental impact of direct emissions. Such offsets provide environmental and economic benefits. The environmental benefits are obvious, in that less emissions means less pollution, and the economic benefits come from the fact that you can use the offsets to lower your overall amount of pollution. Because you are polluting less and probably paying in relation to the amount of pollution, you will have smaller fines, levies, and restrictions. You may even be able to get your emissions low enough to break the threshold where you don't have to pay anything at all.

You can purchase these offsets from many of the same organizations or sources that provide travel and transportation offsets or energy offsets. Because you already know the type and amount of emissions you are discharging, you are ready to contact your offset provider and tell them you want to offset those emissions. They will be able to calculate the carbon equivalent of those emissions so that you can purchase the proper amount of offsets. Once you have determined the amount of offsets required, note it on your worksheet, as shown in Figure 8.10.

When it came to offsetting, Greanco green team members decided that it was not worth it to offset the remaining HVAC emissions from the overall building. However, they were emitting a substantial amount from the forming machine, so they called their offset vendor to determine the cost to offset these emissions. The total cost was substantial, nearly $100 per month, because of the high volume of steel used. Therefore, they decided to offset only 25 percent of the emissions, costing only $25 per month.

Emissions Waste Elimination Worksheet				
Activity or Area:	***Value Stream Activity 1: Forming***			
Future State				
Identify	**Measure**			
Item	Discharge		Savings	Offsetting
	Source	Scrubbing		
Forming machine (steel)	*Wipe off sheets of steel*	*Install scrubber*		*$25 per month (25% of total)*

Figure 8.10 Emissions waste elimination worksheet. Future state: value stream activity 1.

To offset the emissions associated with the end use of your product or service, offer your customers the opportunity to purchase offsets as well. You could do this by setting up your own project to be funded from these purchases, or partner up with a third-party offset provider to offer the offsets to your customers. A great example of this is the airline industry's move to offer their customers the opportunity to offset the environmental impact of their flights.

Step 7: Move toward the Total Elimination of Emissions

This final step is not one that will be achieved overnight; it takes discipline and dedication to completely eliminate emissions. However, achieving this brings with it great rewards. Beyond the obvious environmental ones, it establishes you as an environmental leader in your industry. This is what customers and employees are looking for: organizations that accept the challenge, step up, and develop solutions. They are proud to support an organization that is dedicated to a good cause and that brings with it long-term economic benefits. Also, if you are no longer discharging emissions, you don't have to worry about future compliance, saving you the associated time, effort, and costs. In addition, if you are paying fines or penalties as a result of your emissions, you will no longer be subject to those penalties, saving you those direct costs.

To begin working toward eliminating emissions, start by stepping up your minimization activities. With continued effort, you will have no more emissions from your value stream and overall building. It seems difficult, and

OFFSETTING: WHO'S DOING IT?

A number of different people and organizations are offsetting their emissions in order to do their part for the environment. World Cup Soccer, the Super Bowl, and many other major sporting events are offsetting their emissions. Other entertainment is being offset as well—movie productions and concerts from such artists as the Rolling Stones, Coldplay, the Dave Matthews Band, and Jack Johnson are going carbon neutral. Some large businesses such as Wells Fargo and HSBC are committing to offsetting or going carbon neutral. There are also a number of governments, schools, and churches committing to neutralizing their emissions.

it can be, but by focusing on one item at a time and working through the steps in this chapter to come up with solutions, you will continually improve your organization's ability to tackle this challenge. In addition, you can invest money up-front by investing in technology or specialized pollution-prevention coaches who can assist you along the way. In choosing a coach, make sure that they have a good understanding of what you are trying to achieve and that costs, timelines, and checkpoints for results are set up at the start. You also want to know how they plan to tackle the problem and what previous experience they have that is relevant to your situation. At the end of the day, the most important thing is that you get the results you want. Once you have some ideas on how to move closer to elimination, note them in the Green State section of the Emissions Waste Elimination Worksheet, as shown in Figure 8.11.

CASE STUDY: THE KNOLL GROUP

The Knoll Group manufactures furniture, mainly office furnishings such as desks, credenzas, and chairs. Originally, the company used paints and glues that contained volatile organic compounds. The paint was applied as a spray, leading to emissions of both VOCs and CFCs. The glue also emitted VOCs. Knoll set a goal to eliminate all VOCs from its operations. In looking at the paint, it realized that only 20 percent of the paint sprayed was actually adhering to the furniture, requiring extreme excess of materials and emissions of VOCs. To solve this problem, Knoll looked at powder coating, which turned out to not only eliminate the CFCs and VOCs, but also saved a substantial amount of money in material cost, as 98 percent of the powder-coated paint used stuck to the furniture as opposed to only 20 percent of the sprayed paint. It also used a new hot-glue process that was VOC free and minimized waste. In total, the Knoll group is saving more than $1 million a year from materials, energy, hazardous waste fees, compliance fees, and labor. The cost for new machinery and implementation of the new processes totaled around $1 million as well, providing a simple payback of one year. Each year since, the Knoll group has pocketed $1 million dollars in savings from this initiative alone.

Emissions Waste Elimination Worksheet		
Activity or Area:	***Value Stream Activity 1: Forming***	
Green State		
Elimination		
Item	Solution	Savings
Forming machine (steel)	*Heat crease area only, alterations to machine required*	*Less than 1-yr payback*

Figure 8.11 Emissions waste elimination worksheet. Green state: value stream activity 1.

Greanco green team members are not happy that they are paying to offset the emissions coming from the forming machine. Doing this costs money, and due to a recent downturn in the economy, business has slowed and costs are being squeezed. Because of this, purchasing a scrubber is not possible right now either. The team would rather find a way to eliminate the emissions altogether. Eliminating them would be great marketing material in a time when customers are frugal with their dollars. The team decided to look more deeply into why the steel needed to be heated first. As the members had learned, heating prevented the steel from cracking and made it easier to bend. The team posed questions to engineers to see whether there was any way to get around this process. The engineers discovered that the entire sheet was being heated, but only the part to be bent actually needed to be heated. They were able to make adjustments to the machine so the only area heated was the area where the steel was being bent. This cost money, but the energy savings from not having to heat the entire sheet were substantial and provided a payback of less than one year. This meant that only 30 percent of the sheet needed to be heated and not the entire sheet. Because they were heating 70 percent less than before, emissions dropped by 70 percent as well, moving them that much closer to total elimination. They stopped purchasing offsets for now to save money, and they were better off from an economic and environmental point of view. The sales reps were happy because they now had leverage and differentiation over their competitors.

Chapter 9

The Seventh Green Waste: Biodiversity

What is biodiversity? The answer to this question can be rather complex, as a number of dimensions or aspects are encompassed under the umbrella of biological diversity or *biodiversity*. For the purpose of simplicity, I have chosen a definition of biodiversity offered up by a division of Environment Canada (known as the Canadian Biodiversity Information Network). They define biodiversity as something that:

> ...*encompasses all living species on earth and their relationships to each other. This includes the differences in genes, species, and ecosystems.* Having many different living things allows nature to recover from change. If too much biodiversity is lost, there is a problem because we depend on it to survive. Ecosystems, for instance, are extremely important because they carry out processes such as producing oxygen and cleaning soil and water.

Basically, then, biodiversity is all the things in our environment and how they all relate to each other. It's how the trees, the birds, and the seas all relate to, interact with, and depend on each other to provide an environment that can sustain life. Without healthy biodiversity in an environment, it is impossible to sustain existence.

Biodiversity waste, then, comes from the fact that we pay a price when we take away biodiversity from a particular area and when doing that it has severe and immediate environmental impacts. Biodiversity can be taken away either through one-time destruction as a result of erecting a structure (such as clearing trees or filling in a watershed to put up a building), or through the continuous operation of value stream activities. This may result from actual physical destruction caused by new transportation routes or storing garbage, or it may come in the form of overharvesting. Of course, all the other green wastes contribute to biodiversity destruction in some indirect way, as they all take away some piece of biodiversity. A change in one of these wastes will unquestionably provoke a change in biodiversity somewhere, somehow. Minimizing or eliminating one of the other wastes will help to minimize the specific biodiversity waste it causes.

The reason we have biodiversity as the seventh and final waste is to account for the direct, immediate destruction of biodiversity as a result of constructing the overall building or carrying on with value stream activities. If we did not account for this seventh waste, we would fall short of capturing total environmental impact from the overall building and the value stream. For example, if you have a factory completely powered by solar that harvests and reuses rainwater; reuses its materials in a closed-loop cradle-to-cradle process; has minimal garbage, all of which is recycled; uses only hybrid or biodiesel-powered vehicles and produces no emissions, that's great. But if you then decide to move the factory and build a new plant in the middle of a rain forest and cut down a grove of 100-year-old trees, or if you are a logging company that is overharvesting a forest faster than tress are regenerated, there is definitely an impact on the environment, an impact that may be overlooked if you are not specifically looking for the waste of biodiversity. This is why, by shifting your thinking to look at things from the perspective of the environment and having biodiversity waste as a specific waste you are looking for, will ensure that you have captured your complete environmental impact or ecological footprint.

The end goal is to eliminate the destruction of biodiversity and regenerate what has already been taken. Achieving this will have not only a significant positive environmental impact, but will also have a significant economic impact as well. The economic benefit comes from the savings of not having to destroy that biodiversity in the first place, that is, not paying for tree removal, or to dry out a marsh, or to reshape the land. The savings from this alone are huge. You will also save by not having to pay all the fees now associated with the destruction of biodiversity. In addition, there are economic incentives for not overharvesting. Those incentives come from future revenue you would forgo if you were to overharvest. You will also save on all the red-tape costs associated with permits and applications that are

INTENDING TO REGENERATE BIODIVERSITY

Intending to regenerate biodiversity will fuel you to first find ways to minimize biodiversity destruction so you have less to regenerate. Also, and for the last time, it gives you that ultimate end goal and direction. This gives everything you do in the process meaning and value, as it all contributes toward the end goal. Intending to regenerate biodiversity also shows your commitment to the environment, distinguishing you from the greenwashers out there, and this in itself will help to retain customers and employees, bringing benefits to the bottom line.

required when you take away a piece of biodiversity. There are also rewards in the ultimate end goal of biodiversity waste: that of having a positive environmental impact through regeneration. Obviously, the end goal can take time and effort, which is why we will follow the standardized step-by-step process for waste elimination. The process for biodiversity waste elimination, like the processes for the other wastes, allows you to achieve results and rewards immediately while on the journey toward the end goal:

Step 1: Identify the existence of biodiversity waste either from a one-time activity of putting up a structure or from the continuous operation of value stream activities.
Step 2: Measure the amount of biodiversity waste.
Step 3: Minimize the amount of biodiversity waste.
Step 4: Move toward the regeneration of biodiversity.

Step 1: Identify Biodiversity Waste

To identify biodiversity waste, we are going to break it down into two categories: one-time biodiversity destruction and continual biodiversity destruction. In order to identify one-time destruction, you need to identify all the buildings or structures that contain all the value stream activities plus the support activities outside the value stream (overall building). Remember to include any garages, storage houses, or even uncovered areas such as outdoor inventory holding areas. Also include transport routes that have been installed for your use only (not roads or highways). At some point, each of these areas was home to some biodiversity that has been taken away

Biodiversity Waste Elimination Worksheet		
Activity or Area:	***One-Time Destruction***	
Current State		
Identify	**Measure**	
Item	Type	Amount
Plant and office, parking areas		
Storage shed		

Figure 9.1 Biodiversity waste elimination worksheet. Current state: one-time destruction.

in order for it to function as a storage area, garage, or transport route. To identify these areas, take a walk around the site and note all the structures and holding areas, transport routes, and so on, or talk to the plant, operations, or facilities managers and ask them to help you identify all the buildings/areas. Once you have noted those buildings, structures, and routes, list them in the Current State section on the Biodiversity Waste Elimination Worksheet under the area of one-time destruction, as shown in Figure 9.1.

Now focused on biodiversity waste, the Greanco green team was looking to identify one-time biodiversity destruction. Team members knew that all they needed to do for this step was to identify the buildings, structures, and areas that contained all the activities of their business. Because they had only one location, they took a walk around their property to note what occupied the area. They noted the building that housed the office and plant (including the parking areas), as well as the outside storage shed they used to hold excess materials and finished goods along with tools and other miscellaneous items. They noted these two structures on their worksheet, as shown in Figure 9.1.

To identify continual destruction, you need to look at a few things. First, identify whether harvesting of any natural resources is an ongoing part of your value stream activities. This may include the harvesting of trees, fish, and minerals. If you are directly harvesting any natural resources at all, regardless of how much, note it at this point. If you are harvesting natural resources faster than either you or nature are regenerating them, you are continually taking away from biodiversity. Eventually, there would be nothing left to take, which is unsustainable over the long run; thus, this needs to be identified so it can be dealt with by moving toward the elimination of this waste. In this specific case, that means moving toward sustainable harvesting, in which you are not taking away resources faster than they are being regenerated. To identify continual destruction in the form of harvesting as a waste, observe each value stream activity yourself and see whether any harvesting of natural resources is taking place. If that is not possible, ask the plant manager, operations manager, or the lead hand of that activity if it involves the direct harvesting or destruction of natural resources.

Remember that you are not trying to identify the *use* of natural resources in a value stream activity; that was done in tackling materials waste. And, as we saw in materials waste, it is okay to use natural things as a material as long as those materials are sustainably harvested. If they are not sustainably harvested, you do not have materials waste; your materials are fine since they fit into one of the nutrient cycles (biological nutrient). What you have is biodiversity waste in your supply chain, and your suppliers should be following this process to eliminate that waste. Alternatively, you should choose a different supplier, one that harvests your materials in a sustainable manner, such as wood that is harvested from an FSC-certified forest (which means that a third-party auditor has verified that the forest is managed and harvested in an environmentally and socially friendly manner). If you are doing the harvesting yourself, you are responsible for the biodiversity waste, and you need to follow this process to minimize and then eliminate this waste.

Even if you are not directly harvesting materials as one of the activities of your value stream, you may still be causing continuous biodiversity destruction. It is possible that the activities of your value stream themselves result in the continuous destruction of biodiversity. An example of this is the destruction of grasslands and forests as a result of the continual creation of new transport routes, or possibly the clearing of land if you are a developer, because for a developer, constructing a building is not a one-time situation, it is a continuous activity of a value stream activity. Or, it may come from disposing of garbage or effluents in a natural area that contaminates and destroys habitats of animals, the flow of rivers, and so on. Therefore, you want to identify anywhere that you are continually taking away from biodiversity through the destruction of any part of it. To accomplish this, again, observe each value stream activity to see whether that activity results

Biodiversity Waste Elimination Worksheet		
Activity or Area:	***Continual Destruction***	
Current State		
Identify	**Measure**	
Item	Type	Amount
Item causing continual destruction		

Figure 9.2 Biodiversity waste elimination worksheet. Current state: continual destruction.

in direct, immediate destruction of biodiversity. If you cannot observe the activity yourself, you will have to ask those who are involved with each activity. If you have identified any continuous destruction of biodiversity, whether through harvesting or otherwise, note it on your worksheet under the area of continual destruction, as shown in Figure 9.2.

If you were harvesting a natural resource or continually destroying a natural resource or piece of biodiversity, you would also note that on your worksheet, as shown in Figure 9.2.

Step 2: Measure Biodiversity Destruction

To measure the amount of biodiversity destruction, we are going to continue looking at one-time destruction and continuous destruction in turn. To measure one-time destruction, you need to do two things. The first is to determine *what* was destroyed or taken away, and the second is to measure *how much* was destroyed or taken away. To determine how much was taken away or destroyed, you need to look into what was done to clear the lot for the building or transport route or whatever reason it was cleared. For example, if you are looking at a building, chances are, originally, there was not a perfectly flat piece of solid ground with no trees or grass. Trees had to be cleared, landscape shaped or altered in some way, watersheds filled in, or something of that nature to make it suitable to erect a building. In order to find out what biodiversity was taken away one time in order to erect the building, you will need to talk to the people responsible for taking away the biodiversity. This may be the original owner, the developer, or someone else. If you cannot locate these people, turn to the municipal government that issues building or construction permits, and they should be able to help you determine what biodiversity was taken away in order to erect the building or structure. Failing these two options, turn to historic land records or old photographs to see what the original site looked like and deduce what was done to make it the way it is now.

Knowing what was done to the land or taken away from that site, you can then move on to the next step of measuring how much was taken away. If you determined that originally the lot was treed and those trees were cut down to make a lot suitable for building, you want to determine how many trees were cut down. If a watershed was dried up, how large was the watershed? To determine this, ask the same people who gave you the information on what was done to make it suitable for building; if they do not have exact records of the number of trees or the size of the watershed, they should be able to give you an approximation of the amount. Failing that, you may have to approximate the amount yourself from historic records or photographs. Remember that the point here is to determine, first, what piece of biodiversity was taken away (trees, grass, and watershed), and then how much of it was taken. Once you have determined this amount, note it on your worksheet for one-time destruction, as shown in Figure 9.3.

The Greanco green team was now set to determine how much biodiversity was taken away as a result of the structures being built to house the activities of the overall building and the value stream operations. Team members had no idea where to start; they were the original owners of the building, but no one internally knew the answer. After doing a little searching, they found out who the builder was and contacted them. The builder did not know how much biodiversity was taken away, but was able to provide the green team with the original photograph of the lot on the date it was purchased by the builder. From this picture, the green team could see that there were a variety of trees of different ages originally on the lot but now gone. Team members sent the photograph to a local environmental organization to have them determine the types and ages of the trees in the photograph. This was easily done,

Biodiversity Waste Elimination Worksheet		
Activity or Area:	*One-Time Destruction*	
Current State		
Identify	Measure	
Item	Type	Amount
Plant and office, parking areas	*Oak, maple, cedar trees*	*18, 21, 11*
Storage shed	*Cedar trees*	*6*

Figure 9.3 Biodiversity waste elimination worksheet. Current state: one-time destruction.

and they recorded all the information on their worksheet, as shown in Figure 9.3.

Measuring continuous biodiversity destruction is, unfortunately, highly complicated. A number of variables are involved, such as harvest or destruction rates, growth rates, externalities, max and min thresholds, and different climates, and these all vary from species to species or natural resource to natural resource. To cover even one of these in detail is beyond the scope of this book. However, we will take a generalized look at what needs to be done in order to properly measure what is being taken away as a result of continuous destruction such as harvesting. As with one-time destruction, the point here is to measure what is being taken away.

In the simplest example, if you cut down one ten-year-old tree every day, and plant one ten-year-old tree every day, you would be neutral. However, planting anything less than one ten-year-old tree every day means that you would be taking away from biodiversity. For example, if you plant zero trees, you would be taking one ten-year-old tree per day. If you plant one five-year-old tree per day, you are still taking away from biodiversity, but how much you are taking away depends on a number of variables. Some might say you are planting half of a ten-year-old tree. But does that tree grow at the same rate every year? If not, half a tree is not the answer. What if you plant one ten-year-old tree every day but only 10 percent of the trees you plant survive?

Biodiversity Waste Elimination Worksheet		
Activity or Area:	***Continual Destruction***	
Current State		
Identify	**Measure**	
Item	Type	Amount
Item causing continual destruction	*Biodiversity being destroyed*	*Amount being destroyed per day/week/month*

Figure 9.4 Biodiversity waste elimination worksheet. Current state: continual destruction.

As you can see, even for this simple example, it can get very complicated to determine what is being taken away as a result of continuous destruction or overharvesting. This is why we will not be attempting to cover the details of measuring this source of biodiversity waste in this book, but rather give you the framework for doing it. In order to measure what you are taking away from biodiversity by continuously destroying or harvesting, you will need the expertise of a specialized biologist or scientist. These people will be able to determine the net loss of biodiversity based on your harvest amounts, replacement amounts, and natural regeneration. The good news is that if you are harvesting natural resources or continually destroying biodiversity, you have probably already been through this and have the answer of how much is being taken away. If not, it's very likely that it has been done by government or an environmental organization, so approach these people for help. If this is not the case, then unfortunately you will need to invest in having this done by a qualified specialist. Once you have determined the net loss of biodiversity, note it in your worksheet for continual destruction, as shown in Figure 9.4.

Before moving on to the next step, it is important to note that in measuring the net loss of biodiversity of both one-time and continual destruction, we have looked only at the direct loss of biodiversity and *not* at the indirect losses that are a result of the direct loss. For example, in cutting down a tree, we have not considered the loss of species that lived in that tree. This is something that you should try to measure as well. To do this, consult local environmentalists, government organizations, and so on. You would note this on the worksheet in Figure 9.4 as well.

Step 3: Minimize and Eliminate Biodiversity Destruction

In order to minimize and move toward eliminating biodiversity destruction, we will continue looking at one-time destruction and continual destruction in turn. When it comes to one-time destruction, there is nothing you can do to minimize or eliminate what has already been done. The only way to remedy that impact is to regenerate what you have taken away, which is covered in the last step of biodiversity waste elimination. However, you *can* minimize the future impact of one-time destruction for future projects. There are a number of ways to minimize the amount of biodiversity you are taking away and the impact you are having on the environment in that area.

First, try to occupy an existing building. In this way, you will eliminate any further biodiversity destruction beyond the original destruction of putting up the existing building. You also avoid the generation of other green wastes that would occur if a new building needed to be built. If this is not possible, be very conscious of your site selection. Choose a site that will limit, as much as possible, the biodiversity destruction that will be required to erect your structure. One of the best ways to do this is to build on a brownfield site. A brownfield site is a site that was formerly home to some sort of building or structure but is now considered or perceived to be contaminated. Site-remediation techniques have come a long way in the past few years, and, often, no digging up of the contaminated earth is required. Today there is very little contamination that cannot be cleaned up with some effort. Also, in many cases, governments will subsidize the cost of cleanup and this, coupled with the lower purchase price of the site, can make the overall cost of a brownfield much less than that of a greenfield.

Environmentally, you will have a major positive impact because you have not only avoided further destruction, but also made a positive impact on the environment by cleaning up a site and restoring some biodiversity to it.

If none of these options are possible and you must build on a greenfield site, build on sites that:

- Are not in or near protected areas
- Do not contain protected or endangered species
- Are naturally fit for building in that they require little destruction

Next, when forced to take away a piece of biodiversity in order to build your structure, try to transplant it or build around it instead. Instead of paying to cut down a tree and remove the stump or roots, explore spading the tree and replanting it elsewhere or building around it. If it is a watershed, can you build around or move that by digging a hole beside your lot and pumping the water over, using the earth you dug to fill in the hole where the old watershed was? This has been and is being done in the construction of many buildings to preserve biodiversity while enhancing the aesthetics and performance of the building. Finally, if there is no way to avoid destroying or taking away from biodiversity, try to limit it as much as possible. If you have a choice between destroying a rare or endangered species or another that is more abundant, choose the abundant one. However you plan to minimize or eliminate one-time biodiversity destruction, note the solution or plan in the Future State section of the Biodiversity Waste Elimination Worksheet for one-time destruction, as shown in Figure 9.5.

The Greanco green team spoke with the owners of the company and asked if there were any plans to open up new locations or move the current location. The owners said that they were considering opening up a satellite location if they landed a large customer. After some discussion with the owners, the green team was able to convince them that moving into an existing building rather than building a new one would be the best-case scenario both economically and environmentally. Holding to their commitment to the environment, the owners agreed that should they need a

Biodiversity Waste Elimination Worksheet		
Activity or Area:	***One-Time Destruction***	
Future State		
Minimize		
Item	Solution	Cost/Savings
Plant and office, parking areas	*Future offices to be located in existing buildings*	*Further biodiversity destruction*
Storage shed		

Figure 9.5 Biodiversity waste elimination worksheet. Future state: one-time destruction.

satellite location, they would move into an existing location to limit the amount of biodiversity destruction. The team noted the solution on its worksheet, as shown in Figure 9.5.

Minimizing continual destruction also has substantial benefits. The obvious environmental benefits come from minimizing the amount that is continuously taken away from biodiversity. The economic benefits often come from reduced monetary penalties associated with the destruction of biodiversity, such as fees, permits, and fines. In the eyes of your customers and employees, minimizing biodiversity destruction is seen as a solid commitment to the environment, bringing the benefits of increased loyalty and attraction. Also, in the specific case of harvesting, minimizing and eliminating the continual destruction to biodiversity means that you are moving toward sustainable harvesting. If you are sustainably harvesting, you should be able to sustain the income that comes from harvesting instead of getting a larger cash grab now, and then forgoing future income. This affords more revenue in the long run and increases your value from the perspective of the shareholder. Let's take a look at how we can minimize and move toward the elimination of continual destruction or, in other words, move toward sustainable harvesting or sustainable destruction.

Sustainable biodiversity destruction is all about finding a balance in which you are not taking away biodiversity faster than it is regenerated. As we saw in the Measurement step, this can get rather complicated. We can, however, look at a few general ways of doing this. One way to minimize how much biodiversity you are taking away, either by destroying or by harvesting, is to replace some of what you have taken. This may involve planting trees if you are cutting them down or growing, hatching, and releasing baby fish if you are fishing. Whatever it is you are taking away, replacing some of it will lower the net loss to the environment and minimize biodiversity waste.

If you are already replacing some of what you are taking, another way to minimize continuous destruction is to increase the size or age of what you are replacing. For example, if you are cutting down 10-year-old trees and replanting 1-year-old trees in their place, further minimize your biodiversity waste by planting a 2-, 3-, or 4-year-old tree instead of a 1-year-old. The reason this allows you to minimize is that a 10-year-old tree offers a greater amount of services than a 1-year-old tree. For example, a 10-year-old tree can clean more air and water than a 1-year-old tree can. It is also true that a 2-year-old tree can clean more air than a 1-year-old tree. So if you increase the size or age of the resource or species you are replacing, you are getting closer to fully replacing what you have taken away and have begun to minimize your biodiversity destruction or waste.

Another way to minimize your continuous biodiversity destruction is to reduce the size of your harvest. This one is simple; if you are taking ten trees a year, reduce that to eight or five, and this will begin to minimize what you are taking away from biodiversity. Yet another approach is to lengthen the time between harvesting. For example, instead of harvesting ten trees per year, harvest ten trees every two years. Essentially this gives you a harvest quantity of five trees per year. Doing this definitely minimizes your biodiversity destruction relative to before and will help you get closer to sustainable destruction or harvesting, leading to smaller revenues now but for a longer period of time; in the long run you make more money and have less impact on the environment.

Of course these are very simple suggestions, and before doing any one of these you should consult a specialist. Reintroducing too many, too few, too big, or too little of a species or natural resource may actually hurt more than it helps. For this reason, it is highly recommended for large-scale operations that cause continuous destruction, such as the harvesting of fish and timber, that they consult a qualified specialist in the area. This person will help determine which of these suggestions will work the best. Once you

Biodiversity Waste Elimination Worksheet		
Activity or Area:	***Continual Destruction***	
Future State		
Minimize		
Item	Solution	Cost/Savings
Item causing continual destruction	*Replacing some of what was destroyed or harvested*	*Cost of replacments*
	Increase size or age of replacements	
	Reduce harvest or destruction amount	

Figure 9.6 Biodiversity waste elimination worksheet. Future state: continual destruction.

have figured out how you are going to do this, note the solutions on your worksheet for continual destruction, as shown in Figure 9.6. If you consulted with a professional on the best way to minimize large-scale continuous destruction, the solution would include the actual replacement sizes and quantities.

Move toward the Regeneration of Biodiversity

Regenerating or replacing the biodiversity you have taken away can be quite beneficial both environmentally and economically. From the environmental point of view, you are giving back to nature what you have taken away so that it can continue to provide its natural services such as cleaning the air, water, and soil, which would otherwise have been lost. The economic rewards will come mostly from indirect benefits. These indirect benefits are a result of increased customer and employee loyalty and attraction, a longer-term revenue stream, and, of course, increased value from the perspective of the shareholder. The destruction of biodiversity is oftentimes highly visible, tangible, and public, and therefore the exposure to employees, customers, and shareholders is magnified. In order to regenerate biodiversity, you need to replace what you have destroyed. This is approached differently for one-time destruction and continual destruction, so we will continue to look at each in turn.

Regenerating one-time biodiversity destruction is rather simple. By measuring what you have taken away, your biodiversity waste, you know what you need to replace so that the biodiversity you have destroyed is regenerated. You can replace it all at once or in stages. If you cut down 100 trees to build your building, replace all 100 trees at once, or replace 50 this year and 50 next year, or replace 10 per year for the next ten years. You can approach this in a couple of different ways.

First, plant replacement trees, shrubs, grass, etc., right on your property. However, many times, especially in dense urban areas, there is not enough room to plant or replace what we have taken, so there are other options. A green roof is quickly becoming commonplace on many buildings because not only does it replace biodiversity in the same area from which it was taken, but it also provides energy savings from reduced heating and cooling costs. It is aesthetically appealing and makes your commitment to going green highly visual. Living walls are another way to replace a small amount of biodiversity if you do not have the space. These walls are essentially a vertical garden attached to your building's

CASE STUDY: WHISTLER BLACKCOMB

Whistler Blackcomb is a world-class ski and snowboard resort located in British Columbia, Canada, and will provide three venues for the 2010 Olympic Winter Games. The town of Whistler, alongside the resort of Whistler Blackcomb, has long been dedicated to the environment and achieving sustainability in all its dimensions. Both the town and the resort realize that their livelihood and, in fact, their existence in that area heavily depends on the environment, which is why they want to do everything they can to protect it. The long list of environmental initiatives they have taken includes one involving biodiversity waste minimization. In order to reduce the developmental impacts (one-time destruction) of building a new ski lift in a sensitive subalpine zone, they took a number of actions and precautions. They employed the specialized efforts of biologists and professional foresters to consult on how to minimize their destruction. Also, instead of trekking through the sensitive subalpine regions and lower regions with heavy equipment, they used helicopters to bring equipment in and out to minimize the destruction. In addition to this, they also built the lift when there was snow covering the ground, which acted as a natural protective blanket to the biodiversity underneath. Their efforts all together helped minimize the destruction of biodiversity by reducing the footprint of the operation from an original expectation that it would impact 40 percent of the area, down to only 5 percent.

walls, either inside or outside. The benefits are very similar to that of a green roof.

If none of this is possible, you can replace the biodiversity or an equivalent amount of biodiversity in another area of the world, either by doing it yourself or by contributing to an environmental restoration project or planting project that would not have happened to the same extent if you did not contribute to it. It is important to note here that you should always try to replace the species or resource that you took away originally. If this is not possible, replace it with a native species or resource that requires the least amount of water possible. Introducing new species can negatively affect biodiversity in the area, negating your good-willed efforts. Remember that the end goal is to replace what you have taken, or the natural services that were provided by

Biodiversity Waste Elimination Worksheet		
Activity or Area:	***One-Time Destruction***	
Green State		
Eliminate		
Item	Regeneration	
	Solution	Savings
Plant and office, parking areas	*Plant a combination of oak, maple, and cedar trees at a rate of 11 per year and 12 in the final fifth year*	*56 trees*
Storage shed		

Figure 9.7 Biodiversity waste elimination worksheet. Green state: one-time destruction.

what you have taken, in one way or another. Whatever way you have chosen to do it, note it in the Green State area of the Biodiversity Waste Elimination worksheet for one-time destruction, as shown in Figure 9.7.

Having minimized and essentially eliminated the possibility of future destruction by moving into an existing building, the Greanco green team members were now looking to regenerate what they have already taken away from biodiversity. Because they knew what they had taken away, they developed a plan for replacing it. It was extremely expensive to purchase 10- and 15-year-old trees, so they decided to purchase younger trees and plant them on the lot. Eventually they will grow to replace 100 percent of what was taken away. Their plan was to plant 5-year-old trees, but because these were also expensive, they decided to do it in stages. With 56 trees to replace, they decided to plant 11 trees every Earth Day for the next five years and 12 in the final year. They noted this plan on their worksheet, as shown in Figure 9.7.

When regenerating continuous biodiversity destruction, step up your minimization efforts. In the preceding step, we looked at three ways to minimize the continual taking away or destruction of biodiversity. These three were:

- Start replacing some of what you are taking away.
- Increase the size or age of the species or resources that you are using to replace what you have taken away.
- Reduce the amount that you are taking away or destroying.

In order to fully regenerate the biodiversity you are taking away, you need to step up your minimization efforts so that you are replacing 100 percent. Again, determining the quantity of replacements and the size or age of replacements and the rate at which you replace can get very complicated. Replacing too much at once can be just as big a problem as replacing too little. For this reason, if you are in the situation of harvesting a natural resource or continually taking away or destroying biodiversity, you will need to consult a qualified specialist to determine the best approach to replacing or regenerating 100 percent of what you are taking away. Until you are doing this, you will always be taking away more than is being regenerated, which in the long run is unsustainable. However, if you are not taking away more than is being regenerated, you no longer

Biodiversity Waste Elimination Worksheet		
Activity or Area:	***Continual Destruction***	
Green State		
Eliminate		
Item	Regeneration	
	Solution	Savings
Item causing continual destruction	*Replace 100% of what is being taken away*	*100% of biodiversity destroyed*

Figure 9.8 Biodiversity waste elimination worksheet. Green state: continual destruction.

have biodiversity waste, which means your continual destruction is sustainable in the long run and brings with it sustainable revenue sources. Once you have determined a solution for fully regenerating what you have taken, you would note it on your worksheet for continual destruction, as shown in Figure 9.8. You would include the replacement size, quantity, and rate needed to fully regenerate what you are taking away, as decided by a qualified professional.

CONCLUSION AND APPENDICES

Chapter 10

Conclusion

Change is natural, change is necessary, and change is inevitable. Those who embrace change become leaders of change, and they prosper. Those who resist change become followers and tend to be left by the wayside and forgotten. History has proven this over and over.

The world of commerce is no different. Organizations must continually change in order to develop, grow, and prosper. Changing the way an organization operates to a greener way of being is natural, necessary, and inevitable. This change is going to happen whether you like it or not; in fact, it is already happening. Some organizations have accepted this fact and have embraced changing to a greener way of being; they are quickly becoming leaders in this area. Along the way, the leaders at these companies have been pleasantly surprised to find that making this change is not only good for the environment, but is also profitable, shattering the old paradigm that going green costs money. The new paradigm is green makes you money. This adds more fuel to the fire and speeds up the rate of change to a greener way of being.

The great news is that those who have not yet embraced this change still have time to do so. You can start now, become leaders in this change, and prosper. On the other hand, those who continue to resist this change will eventually be forced to change—by legislation, by competition, or by nature itself. Although a few nongreen companies will survive, most will eventually perish.

In reading this book, you have taken that first crucial step to embracing this change to a greener, more sustainable way of being. You now have a road map to guide you. The introduction to this book showed you that changing to a greener way of being will reduce the negative impacts on the environment to the point that those already-damaging impacts can be reversed, due to the resiliency of this wonderful planet. You also saw through the business case for environmental sustainability that making this change can be done in a way that increases profits either directly (through cost savings) or indirectly (through increased customer attention, employee attraction, and so on).

Chapter 1 provided you with an approach based on one of the world's most successful companies, Toyota. The Toyota Production System, also known as "lean," provides a dynamic system for change, one that is tested, proven, and successful. The only difference between the green value stream approach and a lean approach is merely the fact that the green value stream is focused solely on the environment. Using the green value stream approach to go green will do for you what using the lean approach will do to increase value.

In Chapter 2, you were given the basic road map to carrying out the green value stream process. You saw that, first, a shift in thinking—one that enables you to look at your business from the perspective of the environment—is needed. Getting to know each of the seven green wastes is the next step, after which you can then start to map out your current state by identifying and measuring the seven green wastes in your value stream. Developing solutions to minimize these wastes by following the step-by-step approach for each of the wastes allows you to create your future state and see the opportunities that exist there. Implementing this future green state allows you to realize those opportunities. Moving toward the elimination of each of those seven green wastes creates your green state, at which point you'll see phenomenal benefits. Carrying this methodology across your entire enterprise multiplies those benefits exponentially. Chapter 2 also showed you that, in order to implement a green state within your organization, you have to get management support, and then create a green vision to be carried out under the leadership of a green champion or a team of green champions known as a green team.

Chapters 3 through 9 showed you the detailed steps and the how-to instructions for carrying out those seven steps to create your current state, future state, and green state. To implement this process, it may help to start small and build up momentum. Gain the support of your immediate management and use the green value stream process to pick some of the lower-hanging fruit and show that it works and the benefits that can be realized. You will then gain the support of the top management, be able to create a vision for the entire organization, and be granted the resources to develop green champions to carry out the process in full force.

It is now time to take action, whether you start big or start small. You have all that is necessary to embrace the change toward a greener way of being and become a leader of change to reap tremendous benefits—benefits that extend beyond your organization's bottom line and that are more meaningful, lasting, and important. It's your decision.

Following this chapter are resources that you can turn to for help in following the road map to your green state.

As a closing statement, I will quote Bill McDonough, who once said, "Negligence is knowing better and doing it anyways." After reading this book, you know better. The question I leave with you is: Are you going to keep doing the same things anyways or will you be a leader of change?

Appendix A: Green Dictionary

afforestation: The planting of a forest in an area that previously did not contain forest.

Agenda 21: Released at the Earth Summit in Rio de Janeiro in 1992, Agenda 21 is a United Nations–run program related to sustainable development. It is essentially a plan of action to be taken in order to move globally toward sustainability.

alternative energy: Energy derived from sources other than the traditional sources, such as fossil fuel, coal, and nuclear. Alternative energy includes sources such as solar, wind, hydro, and geothermal.

biodegradable: A type of substance that can be easily broken down by the earth, and once broken down, contributes nutrients back into the earth.

biodiesel: A vegetable-based oil that is used in place of diesel oil. Typically, this oil comes from soy or other vegetable-based products, but can also come from recycled cooking oils.

biodiversity: All the living things on Earth and how they interact and rely on each other in order to create and sustain the different habitats and ecosystems that make up planet Earth. The relationships are rather complex and include the largest creatures and vegetation down to the smallest microorganisms.

biofuels: Environmentally friendly fuels that are derived from renewable sources such as corn, sugar cane, animal waste, and so on.

biosphere: All the places on Earth in which life occurs: land, rocks, water, and air.

brownfields: Abandoned sites, usually old commercial or industrial sites, that have been environmentally contaminated or are perceived to have been contaminated.

Brundtland Report: "The Report of the Brundtland Commission," also known as "Our Common Future," was published by Oxford

University Press in 1987. The report deals with sustainable development and the change of politics needed to achieve such development.

building envelope: All the elements that enclose a building—the windows, roof, and walls as well as the materials contained in them or between them, such as insulations, vapor barriers, and the like.

cap and trade system: A government-legislated system in which the worst offenders or industries are given a cap that limits the amount of **greenhouse gases** that they can discharge into the atmosphere. Those who go over the cap must buy **carbon offsets** or pay a fine to compensate for going over the cap, while those who remain under the cap can make money by being able to sell their excess emission "credits" to those who go over.

carbon dioxide (CO_2): A gas created by the burning of **fossil fuels**, which contain carbon that is the accumulation of millions of years worth of sunlight. Once burned, the carbon is released into the air in the form of carbon dioxide. It is this substance that is the main contributor to the **greenhouse effect** or **global warming**.

carbon footprint: The amount of carbon dioxide that is emitted by a particular entity. It is usually expressed in tons on an annual basis. Oftentimes, this term is confused with ecological or environmental footprints, but these latter categories take into consideration more than just the amount of CO_2 you discharge.

carbon neutral: A net discharge of **carbon dioxide** into the atmosphere equal to zero. Rarely is this achieved purely by not discharging CO_2 into the atmosphere. For the most part, this is done by undertaking projects yourself or purchasing **carbon offset**s that contribute toward projects that will take CO_2 out of the atmosphere, such as planting trees.

carbon offset: Making a financial contribution or "investment" toward a project that will help to remove carbon or, more generally, "greenhouse gases" from the atmosphere. You can offset the environmental impact of everything from your car, to your airplane flight, to your business activities.

carbon sequestration: The practice of pumping CO_2 into the ground in order to keep it from being discharged into the atmosphere and contributing to **global warming**.

chlorofluorocarbon (CFC): A chemical that was previously used in many aerosol sprays and refrigerants. It is no longer produced, but it is still in use in old applications. It is believed that these chemicals produced the infamous hole in the ozone layer.

cogeneration: The process of creating electrical energy that also creates heat, which is then used to create more electricity

compact fluorescent lightbulb (CFL): Lightbulbs that use much less energy than standard incandescent bulbs, resulting in significant savings in energy cost.

crop rotation: Planting different types of crops each season so that the topsoil has a chance to regenerate.

deforestation: The removal of a vast amount of trees in order to make way for roads, buildings, and overall urbanization.

eco-chic: Products or services that are ecofriendly but are also stylish, trendy, and hip.

ecological economics: A field of academia that connects the traditional disciplines of natural and social sciences to study the dependence of human economies on natural ecosystems. This field of study focuses on the issue of how increased human economies will be able to deal with limited and shrinking natural ecosystems.

ecological or environmental footprint: The total amount of ecology or environment (land, water, air, trees, and so on) that is needed to absorb, process, clean, and regenerate the total discharge of pollutants or contaminants and environmental damage caused by a person, business, city, and so on.

ecology: The study of living things in their environment.

ecosystem: The interaction of both the living and nonliving things in a distinct area. A marine ecosystem, for example, refers not only to the fish and the coral that are living in an area, but also the rocks and nonliving things in that distinct area.

embodied energy: The total amount of energy used in the creation of a product or service. This includes the energy used in mining or harvesting, processing, fabricating, and transporting the product.

energy efficiency: Achieving the same results while using less energy.

environmental refugee: A person forcibly displaced due to environmental reasons, such as flooding, natural disasters, pollution, and so on.

environmental sustainability: Operating in a manner that satisfies the following conditions: (1) must not remove materials from the Earth faster than they can be regenerated by the Earth, (2) must not pollute or contaminate the air, land, and water faster than the Earth can process and clean them, and (3) must not destroy or overharvest nature faster than it can regenerate.

erosion: A phenomenon in which soil and land is washed away or "disappears" over time due to constant exposure to the elements of rain, water, wind, and ice.

fair trade: A social movement promoting standards of labor, fair wages, and good employment practices for economically disadvantaged nations.

Forestry Stewardship Council (FSC): A nonprofit organization whose mission is to promote environmentally appropriate, socially beneficial, and economically viable management of the world's forests. FSC-labeled wood products indicate that the wood is harvested from sustainably managed forests.

fossil fuels: Resources buried in the ground, such as oil and natural gas, that have taken millions of years to form. Starting out as living vegetation that was grown with the power of the sun, they have decomposed and, over millions of years, turned into something that we can now use as fuel. This is why they are referred to as the inventory of millions of years of sunlight. They are also considered to be nonrenewable resources.

geothermal power: The generation of energy by utilizing heat stored below the surface of the Earth. Geothermal energy is an environmentally friendly alternative to producing energy, as it has minimal environmental impact.

global warming: The phenomenon of a global increase in temperatures due to the buildup of greenhouse gases (CO_2, NO_2, and so on) in the atmosphere that are trapping excess heat from the sun. Normally this heat from the sun would be reflected back out of the atmosphere. Although the increase seems relatively small, the impact is drastic and potentially catastrophic.

gray water: Wastewater that does not contain human or otherwise toxic waste, such as water that comes from sinks and showers. This water can then be used in situations where potable drinking water is not required, such as the flushing of toilets.

greenfield land: A piece of undeveloped land that is either used for agriculture or is left alone as a natural landscape.

greenhouse effect: As the sun's rays beat down on Earth, about 50 percent of the heat from these rays is absorbed by the Earth; the remainder radiates back into outer space. As the buildup of airborne pollutants traps this heat, more heat is absorbed (instead of radiating back to outer space) than would naturally stay within the Earth's atmosphere. This effect is similar to what happens in a greenhouse to keep it warm, thus the term "greenhouse effect." The pollutants that trap this heat are called **greenhouse gases**. The greenhouse effect leads to **global warming**.

greenhouse gases (GHG): The collection of gases that contribute to a **greenhouse effect** that is warming the Earth by trapping the heat from the sun that would normally have escaped. Greenhouse gases include water vapor, carbon dioxide, methane, nitrous oxide, halogenated fluorocarbons, ozone, perfluorinated carbons, and hydrofluorocarbons. These gases are a result of human activity, such as the burning of **fossil fuels**.

greenwashing: The embellishing of environmental efforts/accomplishments/benefits of a company, product, or service in order to increase sales or reputation. This is dangerous because people who want to reduce their environmental impact may be tricked into purchasing a less environmentally friendly product or service over one that actually delivers the benefits. It is also dangerous for the companies who do this because it is extremely hard to recover from the damage inflicted on their reputation once customers find out.

grid: The network of cables, wires, transformers, and other electrical devices used to transmit electricity from power plants to the end user.

groundwater: Freshwater that is located beneath the ground, usually situated in natural containers known as aquifers. Contaminants and pollutants that leach into the ground can contaminate this water supply.

habitat: The physical area where one or a group of plants or animals live.

heat-island effect: A rise in temperature that occurs in urban areas as a result of replacing natural landscape with roads, buildings, parking lots, and so on. The temperature can be as much as 10° Fahrenheit higher than in the rural areas. This causes an increase in energy costs from increased cooling requirements. You can reduce this effect by bringing back the natural landscape.

kilowatt-hour (kWh): A kilowatt-hour is the standard unit of measure for electricity. One kilowatt-hour is equal to 1,000 watt-hours. It refers to both the amount of power being pulled and the length of time that it is being pulled. For example, 10 × 100-W lightbulbs would pull 1,000 watts, or 1 kW. Turning them on for one hour would consume 1 kWh.

Kyoto Protocol: Implemented in 1997, this initiative is a legally binding agreement, whereby more than 160 countries agreed to reduce their greenhouse gas emissions by an average of 5.2 percent below 1990 emissions levels. It doesn't look like this target will be achieved.

Leadership in Energy and Environmental Design (LEED): LEED is a point-based rating system developed by the U.S. Green Building Council that evaluates the environmental performance from a "whole building" perspective over its life cycle, providing a definitive standard for what constitutes a green building according to six categories: sustainable sites, water efficiency, energy and atmosphere, material resources, indoor environmental quality, innovation, and design process. Buildings evaluated by LEED are rated as: certified, silver, gold, or platinum. There are a total of 69 LEED credits available in the six categories: 26 credits are required to attain the most basic level of LEED certification; 33 to 38 credits are needed for silver; 39 to 51 credits for gold; and 52 to 69 credits for the platinum rating.

life-cycle cost (LCC): The complete cost of purchasing, owning, running, and disposing of something over its entire life span. This includes how much energy or water or natural resources it requires to run it. Energy-efficient or water-efficient products have lower life-cycle costs because, although their up-front or first cost is higher, the cost of operating is much lower due to less energy or water use, and they also tend to last longer.

megawatt: One thousand kilowatts, or 1 million watts; the standard measure of electric power plant generating capacity.

megawatt-hour: One thousand kilowatt-hours, or 1 million watt-hours.

methane: A colorless, odorless, flammable gas forming the major portion of natural gas.

microorganisms: Microscopic organisms so small that you can see them only through a microscope, but that play an integral role in the overall functioning of the Earth as a complete system.

net metering: A billing approach that utility companies use to compensate people for generating their own electricity in excess of the amount they have used. When excess electricity is generated via solar, wind, or some alternative source, it goes back into the grid, causing your hydro meter to actually run backward and, thus, reducing your bill by the excess amount generated. This is a popular approach used by companies that have solar panels.

nitrogen oxides (NO_x): Gases consisting of one molecule of nitrogen and varying numbers of oxygen molecules. This greenhouse gas is a by-product of combustion processes that use fossil fuels and, thus, contributes to the **global warming** phenomenon.

nongovernmental organization (NGO): An organization that operates for the betterment of the environment but is not affiliated with, is not operated by, and does not answer to the government.

nonrenewable resource: Resources that include, but are not limited to, fossil fuels, copper, and nickel that take thousands or millions of years to form, so long that once we use them they will not renew themselves for thousands or millions of years, making them essentially nonrenewable.

organic food: Food grown or raised without the use of chemical fertilizers, pesticides, or drugs. Animals are usually free-range, meaning they are not kept in small cages and are fed a natural diet while being treated benignly.

photosynthesis: The process plants use to change carbon dioxide and water into food by trapping sunlight in chlorophyll.

photovoltaic cell: A device that converts sunlight energy into electricity. This technology has been rapidly advancing and increasingly commercialized over the past few years. It is a fantastic alternative to the traditional fossil fuel approach of producing energy, as it is very ecofriendly. The sun produces enough energy every second to meet current global needs for the next 500,000 years.

postconsumer: Material that was first used by a consumer and now has been turned or recycled into something else.

preconsumer/postindustrial: Material that came from a manufacturing process that has not made its way to a consumer and can no longer be used by the creator.

renewable energy: Energy that comes from sources that naturally replenish themselves in a relatively short time frame. Examples of renewable energy include wind and solar energy.

scrubbers: Equipment attached to smokestacks that is designed to capture **greenhouse gas** emissions before they are discharged into the atmosphere and contribute to **global warming**.

solar cell: See **photovoltaic cell**.

solar power: The power produced by solar or **photovoltaic cells**.

sustainability: Satisfying current human needs without compromising the ability of future generations to satisfy their own needs; operating in a manner that satisfies the following conditions: (1) must not remove materials from the Earth faster than they can be regenerated by the Earth, (2) must not pollute or contaminate the air, land, and water faster than the Earth can process and clean them, (3) must not destroy or overharvest nature faster than it can regenerate itself, and (4) must fairly and equitably cover basic global human needs.

The Natural Step: A framework for sustainability developed in 1989 by Dr. Karl Henrik Robert. This framework gives four system conditions that must be satisfied in order for something to be sustainable. The first three conditions deal with the environment, and the last deals with social and economic issues.

volatile organic compound (VOC): A carbon-based chemical substance that produces noxious fumes and contributes to pollution and contamination. It is found in many paints, caulks, stains, adhesives, and epoxies and oftentimes is exhausted into the atmosphere in mass amounts via industrial production processes.

wind energy: The energy produced as a result of wind turning the blades of wind turbines or windmills. This is the cheapest and fastest-growing type of renewable energy technology.

Appendix B: An Environmental Primer

This appendix provides some brief background information on current environmental issues, their associated impacts, and the idea of sustainability.

What Is Going On?

What is actually happening with the environment today? We hear constantly about global warming, natural disasters, the oil crisis, rain forests disappearing, and species threatened with extinction. We hear about so many different things, and it becomes difficult to get our heads around the whole situation. In order to gain a more holistic perspective and some clarity, let's break down the environmental issues we face today into three major categories or dimensions that align nicely with, and help to articulate, the end goal of sustainability.

The three categories or dimensions are:

- Pollution and contamination
- Core resource depletion
- Direct biodiversity destruction/overharvesting

The environmental issues we face today or that we are threatened with in the future come from one or a combination of these three categories. If we can understand them (what we are doing) and the consequences, we can start to see what we need to do or where we need to go (sustainability), all the while knowing why we need to get there. The challenge then becomes getting there (although with a road map such as the green value stream approach, it can be a rather enjoyable experience). Let's take a look at each of these three dimensions.

Pollution and Contamination

Every second of every day, humans put an enormous amount of toxic substances into the air, land, and water. This pollutes our air and contaminates our water and land, causing major environmental problems. Let's start by examining pollution and one of the largest, or at least most publicized, environmental problems caused by pollution: global warming. Global warming is a result of the phenomenon known as the greenhouse effect. The sun's rays shine on the Earth; some are deflected and some are absorbed. The energy that is absorbed causes the planet to heat up. We do need to absorb some of this energy to be able to live. However, we also release massive amounts of pollution into the air, the worst of which have been infamously termed the "greenhouse gases," which also absorb energy from the sun that would normally be deflected. This causes the planet to heat up more and faster, with multiple consequences.

One major consequence is the effect on ocean currents. There are many different ocean current systems, but they all essentially work like this: Cool water condenses, causing it to fall toward the ocean floor, creating a current or flow. This water travels along the ocean floor until it eventually heats up again and rises to the surface, then continues the flow or current until it cools down and starts the cycle over again. Ocean currents control and determine our climates; if they change, our climates and weather change. When the temperature of the Earth rises, the water does not cool enough to condense and sink. This stops or slows the current or flow and drastically changes climates, weather patterns, and the intensity of storms. Areas that were once hot, humid, and tropical could become cold and dry, and areas that were cold and dry could become hot and humid. This also can cause more intense storms, like Hurricane Katrina. There is no question that, should this continue and increase, it would be disastrous to our economy and to our whole way of life. There are also many other direct consequences to global warming, such as more severe droughts, leading to famine; spread of disease; rising sea levels, leading to coastal flooding; extinction of species; and destruction of entire ecosystems. Pollution also causes more localized, acute impacts such as trouble breathing, or breathing in toxic air that makes us, the animals, and the plants sick, sometimes resulting in death.

We know this is happening; if it continues, we will reach a tipping point after which it will be too late to correct—hence, it is unsustainable. Some argue about how much the temperature or pollution has increased, the rate at which it will increase in the future, or the exact threshold before disaster, but the consequences are generally accepted scientifically, and there is the very real potential for catastrophic disaster if it goes unaddressed. It's like driving down a road, knowing that we are heading for a brick wall in the middle of the road, and arguing about how far ahead the wall is and at what speed we should drive so that we might survive the impact instead of stopping, turning around, and taking a different road.

With a basic understanding of some of the main environmental effects of pollution, most notably global warming, and the consequences associated with these effects, let's now take a look at contamination: the dumping and burying of toxic chemicals in our oceans, lakes, and land. These toxins take the form of anything from plastic bottles to industrial waste to medical waste. They contaminate the land itself, vegetation growing on the land, water below the land, and oceans and lakes, resulting in the death of the life that inhabits that land and water, and the death of the land and water itself in that it will no longer support life. Extinctions occur every day, and an abundance of scientific evidence supports the conclusion that things are dying at a rate that makes regeneration impossible. The problems this causes are manifold. Let's start with the fact that in order to live, we depend on the ecosystems that are being killed by contamination. The entire planet operates as a system, of which we are a part, and when we change this system through contamination, there are consequences. Since we know, for instance, that dumping mercury into a lake kills fish, we can at least start to think about how to prevent that.

Countless studies, costing billions of dollars, have been undertaken to determine how much pollution is in the air; how contaminated oceans, lakes, rivers, and forests are; and what the current and future consequences are. Some studies and numbers are extremely high and some more modest. To arrive at exact numbers is understandably difficult; but there is one solid, indisputable conclusion: there is too much pollution.

Resource Depletion

Every day we increase the amount of resources we pull out of the ground. These natural resources are being harvested at a rate that grossly exceeds the rate at which the Earth can regenerate them, which means that this activity is "unsustainable." The process of extracting resources contributes heavily to pollution, via greenhouse gas emissions, and to contamination, via the discharge of toxins into the land and water. It also directly destroys habitats, resulting in the death of species that live there. Even if we could extract resources with zero environmental impact, all the resources we use to build things, operate our vehicles, and even hydrate ourselves are nonrenewable. These resources represent the inventory of millions of years' worth of energy from sunlight that has driven and enabled the Earth's systems to create them. It's a well-known fact that once we drain the water tables or the oil reserves, they will not fill up anytime soon (essentially, that means "never" in terms of our existence).

In countless reports on how much oil and other resources remain, the information changes constantly. But regardless of exactly how much is left, we know that it is a finite amount, which triggers the laws of supply and demand. The supply of resources is constantly shrinking (a known fact). The law of supply tells us that when we experience a major decline in supply, price increases. Meanwhile, demand for resources is increasing as the population grows and as densely populated countries demand more consumer goods, such as automobiles. The law of demand tells us that when we experience a major increase in demand, price increases. The combination of a major declining shift in supply and a major increasing shift in demand leads to drastic increases in price. Regardless of the exact quantity of resources that are left, we can bank on the price of natural resources continually increasing, and once these natural resources are gone, they're gone.

Destruction and Overharvesting

Next, we will look at the direct destruction of biodiversity and overharvesting.

In order for our society and economies to "evolve," we engage in clear-cutting forests, wetlands, and fields to build more buildings, pave roadways, or grow crops. The problem is that, through the process of photosynthesis, plants, trees, and other vegetation use sunlight to grow and simultaneously provide the services of cleaning water and converting carbon dioxide (CO_2) into oxygen. When large areas of vegetation are destroyed, the Earth's ability to control the amount of CO_2 in the atmosphere and prevent excess absorption of the sun's energy is further diminished. Additionally, entire ecosystems are destroyed, undermining our global ecosystem. Other environmental impacts associated with destruction of biodiversity include erosion, which takes away precious soil we need to grow food, and further contributing to global warming by attracting more of the sun's rays onto pavement and concrete. Another large impact comes from the fact that destroying forests or grazing lands often involves burning, releasing millions of tons of CO_2 into the atmosphere.

Overharvesting land- and water-based resources—taking them away at a rate that grossly exceeds their ability to regenerate—also has significant environmental impacts. First, just as with core natural resources, land- and water-based resources are often harvested in a manner that contributes to pollution and contamination. Second, it contributes to the destruction of the species, habitats, and ecosystems that we are harvesting from. Also, just as with core resources, we encounter the supply-and-demand problem. Even though fish, trees, and plants can regenerate at a much faster rate than core natural resources, if we overharvest, then they cannot regenerate and can be permanently destroyed. For example, fish populations are decreasing, not increasing, and there is a shortage of wood, while the demand for these resources is increasing dramatically. Because of increasing demand and decreasing supply, we can bet that these "renewable" resources will also continue to increase in price. Overharvesting is unsustainable because resources are taken away faster than they can regenerate.

As you can see, the three dimensions of (a) pollution and contamination, (b) core resource depletion, and (c) the destruction of biodiversity/overharvesting are interrelated. Chances are that most of the things that we are doing and consuming today have touched more than one of these dimensions in one way or another, resulting in negative impacts on the environment. We know that if we keep doing things the way we are, we can't sustain life on this planet indefinitely. Since we know there is a wall ahead, we can figure out a different road to take and start moving down that road.

The End Goal: Sustainability

In 1962, Rachel Carson published a book called *Silent Spring* that explored how human-made pollutants threatened to destroy the Earth. This book caught the attention of many people around the world; some call it the birth of the modern-day sustainability movement. Ten years later, in 1972, a meeting in Stockholm focused on a report called "The Limits to Growth," which identified the fact that there are limits to the amount of growth the human race could experience if we keep doing things the way they were being done. This opened many people's eyes to this major problem. Fifteen years later, in 1987, the Brundtland Report (also known as "Our Common Future" report) was issued. This report basically deals with sustainable development and the political changes needed to achieve it. The famous definition of sustainability also appeared in this document: "Sustainable development is development that meets the needs of the present without compromising the ability of future generations to meet their own needs." Also in 1987, the Montreal Protocol was developed to deal with the ozone problem. Two years later, Dr. Karl Henrik-Robert developed The Natural Step system, a development that allowed the sustainability movement to gain momentum as it spawned many of the current ideas, concepts, and techniques associated with sustainability today. Then, in 1992, another meeting in Rio de Janeiro produced "Agenda 21," a comprehensive plan of action to be taken from a global to a local level to move toward sustainability. The year 1997 marks the development of the now-infamous Kyoto Protocol, an agreement for industrialized countries to lower their greenhouse gases (GHGs) to 5 percent lower than they were in 1990. Between then and now, there has been a tremendous growth in the size and momentum of the sustainability movement.

With that snapshot of the history of the sustainability movement in mind, let's take a look at what sustainability is. There are a number of definitions, concepts, and ideas surrounding the topic. Does it all have to do with the environment? What about the social and economic aspects? In order to clear up the confusion and provide a clear understanding of what sustainability is, I use The Natural Step system developed by Dr. Karl Henrik-Robert. At the core of The Natural Step philosophy is the idea that, in order for our planet to be sustainable, that is, to be able to continue the existence of human and other life, we must satisfy four system conditions:

1. Substances from the Earth's crust must not systematically increase in the biosphere. In other words, we cannot extract core resources at a faster rate than they can go back into the Earth and regenerate. If we do, eventually there will be nothing left, making it unsustainable.
2. Substances produced by society must not systematically increase in the biosphere. In other words, we cannot pollute and contaminate at a rate faster than the Earth can process and cleanse the pollutants and contaminants. If we do, we will get to the point where there is not enough air and oxygen needed for life; again, this is unsustainable.
3. Nature's functions and diversity must not be systematically impoverished by physical displacement, overharvesting, or other forms of manipulation. In other words, we cannot destroy or overharvest nature

faster than it can regenerate itself. If we do, eventually it will all be destroyed; this is also unsustainable.
4. Resources must be used fairly and efficiently in order to meet basic human needs globally.

Three of these conditions deal with the environment, and one applies to society and the economy. The first three system conditions that deal with the environment are negative; they state what we can't do. This is because sustainability can be so many things and take so many forms that it would be almost impossible to list all of them. It is much easier to say what sustainability is not, and then take action to ensure that what we do satisfies those conditions. The three dimensions of environmental impact introduced above also align very nicely with the first three system conditions. If we abide by the first three system conditions, we would deal with the three dimensions of environmental impact.

Sustainability, therefore, means acting in alignment with and completely satisfying each of the four system conditions. If everything we do satisfies these four conditions, we can say that we are sustainable; anything else means that we are moving *toward* sustainability. So, if we do not pull core resources from the ground faster than they are regenerated; if we do not discharge toxins into the air, land, and water faster than the Earth can process and cleanse them; if we do not destroy or overharvest nature faster than it can regenerate; and if we distribute resources fairly and efficiently so that basic human needs are met globally, then that is sustainability.

We can also deduce another definition from this: the definition of what it means to go green. To go green means to achieve environmental sustainability or, in other words, to satisfy the first three system conditions. So, if we are operating in a way whereby we are not pulling core resources faster than they are regenerated, we are not discharging pollutants and contaminants faster than the Earth can process and cleanse them, and we are not destroying or overharvesting nature faster than it can regenerate what we have taken, then that is environmental sustainability. In other words, that is truly "being green." Anything short of that means that we are "going green," or on the way to being green. Because this book focuses on going green or environmental sustainability, we focus on satisfying the first three system conditions only. However, by satisfying the first three system conditions, we will inevitably have an impact on the social and economic aspects of system condition 4.

Appendix C: Resources

The journey toward environmental sustainability, although challenging, should be enjoyable, exciting, and rewarding. However, time spent in reinventing the wheel or searching for answers in the wrong place is an exercise in frustration. To help you avoid such frustration—and in keeping with the theme of simplicity and making the green journey as easy and beneficial as possible—you will find below a number of resources that will help you on your journey toward sustainability. This appendix is meant to aid you as you work through each of the chapters. It is organized by chapter, so if you come across a step or idea in a chapter that you need help in further understanding, or even help in getting started, you can turn to this appendix for support. You can also visit www.greenenterprise.ca for further resources, support, and companion materials to this book, such as the free download of the worksheets.

General Environmental and Overall Sustainability Resources

The references in this section will help you to gain a better understanding of the current environmental issues, the end goal of sustainability, and the organizations dedicated to creating a sustainable future.

www.greenenterprise.ca: Visit this Web site for more information on the GVS approach to book a green value stream and training workshop or for access to a number of green resources.

www.globalreporting.org: The Global Reporting Initiative is the world's leading reporting framework for organizations to report their performance on sustainability.

www.davidsuzuki.org: David Suzuki's foundation, which is dedicated to developing and promoting ways that society can sustainably live in balance with the natural world. There are educational resources, the latest news regarding environmental issues, and other resources to bring and keep you up to speed on the latest environmental issues.

www.paulhawken.com: Paul Hawken is one of the world's leading environmentalists. He is also an entrepreneur and author of six best-selling books. His Web site will introduce you to recommended further readings as well as the organizations that he is involved with. Much of his work has been "game changing" stuff, and becoming familiar with it is highly recommended.

www.naturalcapital.org: The Natural Capital Institute is one of the organizations that Paul Hawken is involved with. In their own words, they "serve the people who are transforming the world."

www.naturalstep.org: The Natural Step is an organization that many believe to be one of the pioneers of modern-day sustainability. Reference this Web site for some great case studies as well as a number of educational resources and tools to help you in your journey toward sustainability.

www.epa.gov: The U.S. Environmental Protection Agency Web site. There is so much information regarding the environment that it is hard to summarize: everything from tips on going green to the latest in the development of environmental legislation.

www.bigpicture.tv: Big Picture TV streams free videos from the leading thinkers and doers in the world of sustainability. You can join as a member and download the videos to use in presentations.

www.planetfriendly.net: Environmental organization dedicated to providing the latest information regarding environmental jobs, volunteer opportunities, and environmental events in your region.

www.ted.com: TED (technology, entertainment, and design) is an organization that brings forth talks by the world's leading thinkers and doers. Centered around their annual conference in Long Beach that is packed with talks from these leading individuals, they now make these talks available for free to inspire others to make a difference.

www.nrcan.gc.ca: Natural Resources Canada Web site, full of information on a wide array of topics such as earth sciences, mining, forestry, and (of course) energy-related issues. A good starting point to dive deeper into the increasingly important topic of natural resources.

www.ec.gc.ca: Environment Canada's official Web site. This is a great portal to all the different environmental programs offered by the government of Canada: everything from grant and incentive programs to educational resources.

www.cagbc.org: The Canadian Green Building Council is responsible for administering the LEED certification program in Canada. If you are looking to certify your building, to become an accredited LEED professional, or for ways to green your building, this is a great resource.

www.usgbc.org: The U.S. Green Building Council is responsible for administering the LEED certification program in the United States. If you are looking to certify your building, to become an accredited LEED professional, or for ways to green your building, this is a great resource.

www.wwf.org: World Wildlife Fund's Web site. Dedicated to the conservation and preservation of wildlife around the world. If you're looking for ways to get involved in your community or general information related to the current environmental issues affecting wildlife around the world, this is a great resource with a long-standing reputation of actually making a difference.

www.sustainability.com: A sustainability think tank that is devoted to providing business and other organizations with the risks and opportunities that come along with sustainable development

www.unesco.org: The United Nations Education, Scientific, and Cultural Organization was founded in 1945 and yields a tremendous amount of ability to progress the movement toward sustainable development.

www.unep.org: The United Nations Environment Program is the voice for the environment within the United Nations. There is a wealth of information on this Web site of everything from educational resources and tools to specific ways to get involved and make a difference.

www.sam-group.com: The Sustainable Asset Management Group works with the Dow Jones Sustainability Index in doing research activities and helping to identify the world's leading sustainability-driven companies. Their studies, research, and reports are an important part of the growing interest in investing in sustainability-driven companies.

www.sustainability-indexes.com: The Dow Jones Sustainability Index tracks the financial performance of the world's leading sustainability-driven companies.

www.ecoeco.org: The International Society for Ecological Economics Web site. Their goal is to promote cooperation and understanding between ecologists and economists to advance sustainable development. The field of ecological economics pulls from many different fields of academia to make sense of the issues surrounding population with shrinking environmental resources.

www.rmi.org: The Rocky Mountain Institute is a think tank for environmental- and sustainability-related issues. Founded by Amory Lovins, this organization can answer or help with many of your questions or issues related to the environmental challenges you may be facing.

www.grist.org: A nonprofit, green news organization. They are dedicated to bringing you the latest and most up-to-date news in the world of environmental sustainability. Read news and articles on everything from biofuels to polar bears. This is one of the world's largest providers of trusted environmental news.

www.greendaily.com: A leading environmental blog dedicated to providing insights and discussions on the latest environmental topics and issues.

www.environmentaldefence.ca: A Canadian NGO helping decision makers and businesses bring environmental issues to the forefront. They champion a number of environmental campaigns and offer educational resources such as free downloads of published environmental reports and studies.

www.acee-ceaa.gc.ca/index_e.htm: The Canadian Environmental Assessment Agency is a branch of the Canadian government, helping organizations make informed decisions about environmental issues.

www.iisd.org: The International Institute for Sustainable Development is committed to helping organizations move toward a world of sustainable development. Includes a publications center with the latest articles, books, and papers.

www.nrcan-rncan.gc.ca/sd-dd/index_e.html: Natural Resources Canada Sustainable Development Web site provides up-to-date information regarding Canada's sustainability policies as well as a wealth of information, tips, best practices, case studies, and so on, to help your organization move toward sustainable development.

www.earth-policy.org: The Earth Policy Institute's Web site. The EPS is headed up by Lester Brown, a leading environmentalist, author, and advocate for a sustainable economy. A resource that contains a plethora of articles and papers relating to a sustainable economy and ways of achieving it.

www.billmckibben.com: Bill McKibben is an author and educator in the field of environmentalism. One of the giants in this field, McKibben offers up a great deal of insight on how we can personally make a difference.

www.edf.org/home.cfm: The Environmental Defense Fund partners up with organizations of all levels to help them find solutions that move toward sustainable development. This group is a major player in the environmental arena.

http://green.msn.com: MSN's green site. This is a great resource bringing you feeds of the latest environmental news, practical tips, and interesting stats relating to environmental issues.

www.treehugger.com: Possibly the world's leading environmental blog. You will find information here on everything from greening your business to greening your personal life. Includes tons of videos, articles, and discussions all relating to the environment. Definitely one to put in the favorites list.

Chapter 1

In this chapter, we first looked at how the lean philosophy provides the framework or structure for supporting the overall green value stream (GVS) approach. The GVS approach is designed to be used as a stand-alone approach, but it can also be used, very successfully, in tandem with lean efforts. If you would like to learn more about lean and the value it can add to your organization, check out the following resources:

www.lean.org: Possibly the most trusted and useful resource for lean information, this is Jim Womak's nonprofit organization dedicated to helping organizations become leaner. It contains articles, a library of books, workshops, and so on.

www.ame.org: Association for Manufacturing Excellence is dedicated to helping manufacturers move toward operational excellence and has some great resources for those looking to educate themselves on lean, to network with those who are already doing it, or to be exposed to the latest in manufacturing excellence.

www.cme-mec.ca: Canadian Manufacturers and Exporters Association is dedicated to helping Canadian manufacturers and exporters succeed in an increasingly competitive global market. From lobbying government for manufacturers' rights, to holding information sessions on everything from health and safety to doing business in other parts of the world, to networking with some of Canada's top business people and holding Lean 101 workshops, these people are here to help in a number of ways.

If you are looking for help or coaching in implementing lean within your organization, see the following resources:

www.hpsinc.ca: High Performance Solutions, Inc., is Canada's leading lean consulting firm. Their approach is to teach you and empower you to be able to carry out lean activities on your own without their help, an approach you very rarely see from consultants. When you are looking for help in your lean efforts, you can count on the most trusted consultants in the industry today.

www.hpmconsortium.com: The high-performance manufacturing consortium is a lean manufacturing consortium bringing together a small group of companies that share best practices and offer training sessions in order to compete and win at the highest level. Visit its sister consortium at www.afee.ca.

www.productivitypress.com: For industry's largest collection of organizational improvement books and materials, this group is a great resource when first starting your journey, well into your journey, or just keeping up with the latest material in organizational improvement.

Chapter 2

Chapter 2 looked at the process for carrying out the green value stream approach. We also looked at how you would go forth and implement the green value stream approach in your organization. The following resource helps you carry out the green value stream approach and implement it in your organization:

www.greenenterprise.ca: A coaching firm dedicated to educating, training, and coaching you on the green value stream approach so that you can implement this approach in your organization. Also contains further resources to help you in your green journey and companion material to this book.

Chapter 3

In this chapter, covering the first of the seven green wastes, we looked at what the waste of energy is and reviewed the steps to minimizing and eventually eliminating this waste. You can turn to the following resources for help in the first couple of steps of identifying and measuring energy usage as well as general information related to energy issues:

www.energy.gov: U.S. Department of Energy Web site that contains a slew of information relating to energy, energy efficiency, biofuels, the latest issues, government policies, incentive programs, and a number of energy-related topics.

www.iea.org: International Energy Association Web site. The IEA is an energy policy advisor to 27 member countries and advises these countries on everything from energy security to environmental issues. They produce a large number of reports and studies that are useful in getting up to speed on global energy issues.

www.centreforenergy.com: Canadian Energy Information Web site provides a number of great tools for getting up to speed on energy issues and the energy market in Canada. It provides insight into the market as well as a number of educational resources for everyone from K–12 students to business leaders.

www.aeecenter.org: Association for Energy Engineers is a nonprofit professional society providing professionals in all areas related to energy management, measuring, and planning. They also provide training courses and certifications related to all things "energy."

www1.eere.energy.gov/consumer/calculators: The U.S. Department of Energy's various energy-related calculators can help you in determining your carbon footprint and overall environmental footprint.

www.ieso.ca: Ontario's Independent Electricity System Operator. This group balances the supply and demand of energy in Ontario. A great resource for energy-related items, such as determining how you are billed as well as tips for identifying, measuring, reducing, and managing your energy consumption.

www.ashrae.org: American Society of Heating, Refrigerating, and Air Conditioning Engineers Web site. This organization is a great resource and plays a big role in determining energy efficiency standards for the green building industry. They offer a number of publications, educational tools, and certifications regarding energy, or more specifically, HVAC's contribution to energy consumption.

www.tcf-fca.ca/calculator: A carbon emissions calculator.

www.cleanerandgreener.org/resources/calculators.htm: Another carbon emissions calculator.

For help in minimizing the amount of energy waste you have in your organization or value stream, you can turn to the following resources for help:

www.energystar.gov: U.S. Energy Star Web site. We see this logo everywhere, from our fridge to our computers. This Web site will help you to locate energy-efficient products for everything from lighting to computers to HVAC components. They also provide educational resources and other tools, such as typical payback periods and calculators for determining specific paybacks. They also determine the standards for energy-efficiency levels and govern those standards.

www.oee.nrcan.gc.ca: Canada's Office of Energy Efficiency is a government-run office that provides education and resources for minimizing your energy consumption. They also offer a number of incentive programs for commercial and industrial organizations.

www.ecologo.org: Canada's Eco Logo Program is a green product certification program. A number of their certified products, although not all, will help you in your quest to minimize energy consumption in your organization.

www.usgbc.org: U.S. Green Building Council administers the LEED certification program in the United States. The leading green building certification program in the world, the council offers a wide array of tips, techniques, and tricks for reducing energy consumption from your building-level systems such as HVAC and lighting. You can find a lot of these tips in their reference packages.

www.cagbc.org: Canada Green Building Council administers the LEED certification program in Canada. Just like its U.S. counterpart, the Canadian organization offers a wide array of tips, techniques, and tricks for reducing energy consumption from your building-level systems such as HVAC and lighting. You can find a lot of these tips in their reference packages.

www.iesna.org: Illuminating Engineering Society of North America is the recognized technical authority on lighting. You can refer to them for ideas, tips, and technical information regarding the latest and greatest ways to conserve energy through lighting retrofits.

Finally, for help in moving toward the elimination of energy use and providing your own self-generated clean energy, you can turn to the following resources to help you get started and find more information:

www.re-energy.ca: Renewable Energy Web site, which provides a wealth of educational information regarding renewable energy. They give you everything from the overall basics of renewable energy to the specifics of each form of renewable energy, such as solar, wind, and biomass.

www.canhydropower.org: The Canadian Hydropower Association Web site. Use this resource to contact the major players in the hydropower industry as well as to learn more specifically about hydropower.

www.bullfrogpower.com: Bullfrog Power is a green power provider. This is an excellent alternative to purchasing carbon-offset credits once you have minimized your consumption of energy. This group is *the* green power provider in Canada.

www.pembina.org: The Pembina Institute's mission is to advance sustainable energy solutions. This group advocates, educates, and raises awareness around renewable or sustainable energy solutions. A great launching pad to find more in-depth, detailed information on specific solutions.

www.energyalternatives.ca: Renewable Energy Solutions is a private company that has installed thousands of renewable energy systems.

Chapter 4

In this chapter on water, we saw that water waste comes from the fact that we are paying someone else to use excess amounts of water, paying to contaminate it, and paying again to discharge it, along with all the attendant environmental impacts. Instead, we could be using only the water we need, harvesting some of it for free, and then working toward cleaning it ourselves and reusing it. For help in identifying and measuring water waste and general information on the issues surrounding water, here are some resources to get you started:

www.epa.gov/OW: EPA's water Web site dedicated to all things water, from water science to water education and conservation. This Web site is a good resource for getting your feet wet with water.

www.water-ed.org: The Water Education Foundation Web site. The goal of this nonprofit organization is to provide a better understanding of water issues and help solve water resource problems.

www.nwri.ca: National Water Research Institute of Canada. This is Canada's freshwater research institute. With research sites across Canada, these people are the experts in linking water issues to environmental policy.

www.on.ec.gc.ca/reseau/waterCalculator/login_e.html: Government of Canada's water use calculator.

www.standupfortheearth.org/green-tools-education/roi-calculator: A water usage calculator that helps to determine the ROI on water reduction investments. Provides typical payback scenario for an office building.

For help in minimizing the waste of water you can turn to the following resources:

www.irrigation.org: The Irrigation Association Web site. This organization is a member-based organization that also serves to define and educate on best practices of water management.

www.americanwater.com: America's oldest and largest water conservation company. Experts at tapping into the savings available from water conservation. Contains a list of nearly 100 water-saving tips that you can use both at home and at work.

If you are looking for help in harvesting rainwater, check out the following resources:

www.arcsa.org: American Rainwater Catchment Systems Association Web site. This organization is dedicated to the promotion of rainwater harvesting. They offer educational resources, training, and workshops as well as forums for discussions on the latest techniques and technologies relating to rainwater harvesting.

For more information on the reuse of water by cleaning water yourself and moving toward the elimination of this waste, the following resources will give you a better understanding of what is involved and help get you going in the right direction.

www.awwa.org: American Water Works Association Web site. This organization is over 125 years old and has a wealth of information regarding all things water. With over 60,000 members, it is the largest association of water professionals in the world.

www.thewaterhole.ca: This resource opens the floodgates when it comes to issues surrounding the protection of water resources. They can help you get up to speed on the latest water-related issues, debates, and events.

Chapter 5

Materials waste is one of the greatest sources of environmental pollution, and we saw in this chapter that its management provides a significant opportunity for savings and benefits beyond the environmental ones. Check out the following general resources on this waste as well as additional resources that will help you start to move on through the steps of identifying, classifying, and assessing materials:

www.mbdc.com: Web site for McDonough/Braungart Design Chemistry. This is the Web site for all things cradle-to-cradle; these people invented it, and no one knows it better or can provide you with more information regarding the cradle-to-cradle process. This is also the Web site to consult if you are looking for cradle-to-cradle certification for your products.

For help and support in minimizing materials use and phasing out the bad materials, here are some resources to help you get started with these steps:

www.hazmatmag.com: A resource that will bring you up to speed on the latest issues surrounding hazardous materials. There are many articles, papers, and discussions that can help you to deal with your hazardous material issues.

www.basel.int: The Basel Convention's Web site. The Basel Convention is the leading source on international movement and disposal of hazardous waste.

www.greencleaninstitute.com: The Green Clean Institute provides certification of green janitorial services.

www.greenseal.org/about/index.cfm: Green Seal is a nonprofit organization that provides certification to products that meet minimum environmental requirements.

We saw that having the ability to reuse materials has to start at the design stage, so if you are ready to start moving toward the elimination of this waste and reusing materials, here are some places you can turn to for help and inspiration:

www.owe.org: Ontario Waste Materials Exchange Web site. This program is designed to facilitate the reuse or recycling of postindustrial waste. You can post your waste for someone else to buy or check out whether there is something that you can use.

www.scrapmatchga.org: Another materials-exchange Web site, this one located in Georgia. These programs are great for finding a new home for your garbage so that it does not end up in the landfill and can reduce waste-removal costs, maybe even providing a revenue stream.

www.epa.gov/dfe: EPA's Design for Environment Web site is committed to helping you design products that are good for the environment.

www.buildingreuse.org: The Building Materials Reuse Association Web site. This nonprofit association is dedicated to facilitating the reuse and recycling of building materials. They also aim to educate the public on the benefits of this increasingly popular practice.

Chapter 6

We saw in this chapter that the waste of garbage is a double-edged sword, costing you up-front to purchase excess materials, and then costing on the back end to have it taken away, so this is a waste where you can quickly reap some savings. Check out the following resources that can help you in identifying, measuring, and minimizing the amount of garbage waste you have, allowing you to move toward the elimination of this waste:

www.turtleislandrecycling.com: Turtle Island is a recycling company that specializes in diverting your waste from the landfill. They have the ability to recycle or reuse many different materials and are the leader in multimaterial recycling initiatives.

www.recycle.nrcan.gc.ca: Natural Resources Canada's recycling Web site. Provides a wealth of information regarding what can and can't be recycled, the various recycling processes, and a database of the companies that can recycle a variety of materials and minerals.

www.rcbc.bc.ca: The Web site for the Recycling Council of British Columbia. This nonprofit organization promotes the principles of zero waste and offers various tips, ideas, and suggestions on how to help you minimize your garbage.

www.rco.on.ca: Recycling Council of Ontario Web site. This nonprofit organization is committed to environmental sustainability by doing its part to eliminate waste. Check out this site for a vast amount of resources related to the three Rs.

Chapter 7

After going through this chapter, we saw that travel and transportation waste is rampant in most organizations; yet it seems to go unnoticed and continually increases. To get started toward minimizing this waste, check out the following resources:

www.fueleconomy.gov: The U.S. Department of Energy's fuel economy Web site. A great resource for not only getting tips about improving your gas mileage, but also learning about alternative transportation.

www.terrapass.com: Another provider of carbon offsets, they also provide carbon-footprint calculators and offer other products and services to reduce that footprint.

Once you are ready to move toward green or eco-transportation, you can turn to the following resources:

www.greenercars.com: This Web site compares the ecofriendliness of a number of different vehicles on the market. You can use this resource when you have made the decision to move toward eco-transport and are deciding on which vehicle is best for your organization.

www.atti-info.org: Web site of the Advanced Transportation Technology Institute. This organization is on the leading edge of alternative-fuel vehicles that emit low or zero emissions. They can assist your organization in determining and implementing an ecofriendly transportation solution that will work best for you.

www.consumerenergycenter.org/transportation/consumer_tips/index.html: A California-based organization that helps bring you up to speed on environmentally friendly transportation options.

www.cta-otc.gc.ca/index_e.html: The Canadian Transportation Agency Web site. It provides news and information related to the issues and policies surrounding travel and transportation in Canada.

www.tc.gc.ca/programs/environment/UTEC/menu-eng.htm: Transport Canada's emissions calculator.

www.carbonzero.ca/calculator/index.php: An offset company with a great travel and transportation emissions calculator. No commitment to buy for using their calculator.

Chapter 8

Although we know that the other wastes contribute to emissions and effluents remotely, we saw in this chapter that waste in the form of emissions and effluents also comes from the direct creation of these emissions on-site. With government legislation increasingly focusing on this type of waste, and with enforcement just around the corner, you can bet you'll be tackling this waste shortly, whether it is voluntary or not. To get a head start on moving toward the minimization and elimination of this waste, see the following resources:

www.p2pays.org: A pollution-prevention Web site supported by the North Carolina Department of Environment and Natural Resources. This site is packed with information regarding pollution-prevention solutions—everything from best-practice case studies to technical and compliance assistance.

www.epa.gov/p2: The EPA's pollution-prevention Web site is full of great tips for both the beginner and the advanced pollution preventionist. Whether you are looking for basic information or strategies regarding different grants and funding or policies surrounding the issue, you can find it here.

www.c2p2online.com: Canadian Center for Pollution Prevention is an organization that encourages and supports positive environmental action through the sharing of knowledge and tools. They provide conferences and training, a variety of tools and resources, and networking opportunities with others facing the same challenges.

www.inece.org: The International Network for Environmental Compliance and Enforcement Web site. They are dedicated to providing awareness surrounding environmental compliance.

www.p2.org/index.cfm: The U.S. National Pollution Prevention Roundtable. Dedicated to everything involving pollution prevention. A great resource for digging into the issues surrounding this waste and identifying solutions to eliminate it.

www.mcex.ca/index_en: The Montreal Climate Exchange Web site. An open-market vehicle for carbon (emissions) trading in Canada.

www.chicagoclimatex.com: The Chicago Climate Exchange Web site. An open-market vehicle for carbon (emissions) trading in the United States.

Chapter 9

We saw that the biodiversity waste is difficult to identify as a source of hard, direct cost savings, but the benefits in terms of employee and customer loyalty and attraction are significant. For help in moving toward the elimination of this waste and to see how others are leveraging their efforts, check out the following resources:

www.cbin.ec.gc.ca: The Canadian Biodiversity Information Network Web site. This site is a portal into the world of Canadian biodiversity. Get access to the national biodiversity strategy, information on biodiversity, or tips on things that you can do to protect the biodiversity in your area.

www.biodiversityhotspots.org: A Web site supported by Conservation International, this site lists the world's richest but most threatened areas of plant and animal life. This site is a good resource if you are looking to get up to speed on the global issues of biodiversity.

www.ser.org: More than 20 years old, The Society for Ecological Restoration is a nonprofit organization dedicated to the restoration of ecosystems throughout the world. They do not do the restoration themselves, but rather facilitate and help others to gain the knowledge and ability to implement restoration projects on their own.

www.plantnative.org: This organization is dedicated to moving native plants into the mainstream of landscaping practices. This site has a vast amount of information, from finding out what plants are native to your area to different books, directories, and how-tos on making native plants part of your landscaping.

www.nanps.org: The North American Native Plant Society is dedicated to the conservation of native plants. This organization, almost 25 years old, can help you with any questions you may have about native plants and the issues that surround this important topic.

www.wildflower.org: This organization was started Lady Bird Johnson, an entrepreneur, journalist, and former first lady of the United States. Like the two resources directly above, this organization is dedicated to conservation and sustainability of native plants.

www.greenroofs.org: A green-roof association dedicated to educating the public on the how-tos, as well as the benefits, of putting in a green roof. They offer a number of educational courses, green-roof calculators, and an accreditation program in addition to organizing an annual conference on green roofing.

www.aboutremediation.com: A great resource for learning about site remediation and reclamation. The Web site includes lots of information regarding the remediation process, the financial opportunities that exist in remediation, and even best-practice case studies as well as a toolbox.

http://epa.gov/cleanup: The EPA's remediation and reclamation Web site. A great place to start when looking into or being forced into cleaning up a site that has been contaminated.

Appendix D: Worksheets

Energy Waste Elimination Worksheet

Activity or Area:					
Current State					
Identify		*Measure*			
Item	*Source*	*Rate*	*Usage*	*Consumption*	
				Qty	*Cost*

Future State						
Minimization						
Item	*Conservation*		*Efficiency*		*Management*	
	Solution	*Savings*	*Solution*	*Savings*	*Solution*	*Savings*

Green State				
Elimination				
Item	*Offsetting*	*Renewable Energy*		
		Solution	*Cost*	*Savings*

Figure 14.1 Energy waste elimination worksheet.

Water Waste Elimination Worksheet

Activity or Area:	

Current State

Identify	Measure				
Item	*Flow Rate*	*Usage*	*Consumption*	*Discharge*	*Toxicity*

Future State

Minimization

Item	*Quantity*	*Toxicity*	*Savings*

Green State

Elimination

Item	*Rainwater* (insert harvest amount)			*Re-use*	
	Filtered?	*Supply RW?*	*Savings*	*Grade*	*Use*

Figure 14.2 Water waste elimination worksheet.

Materials Waste Elimination Worksheet

Activity or Area:	

Current State

Identify			*Measure*					
Input		*Output*	*Material Makeup*		*Classify*			*Assess*
Item & Quantity	*Material Makeup*	*Item*	*Input*	*Output*	*Tn*	*Bn*	*LF*	

Future State

Minimize

Item	*Harmful Materials (Phase Out)*	*Quantity*	*Savings*

Green State

Eliminate

Item	*100% Tn & Bn Inputs*	*100% Tn & Bn Outputs*	*Re-Use*	*Savings*

Figure 14.3 Material Waste Elimination Worksheet

Garbage Waste Elimination Worksheet			
Activity or Area:			
Current State			
Identify	*Measure*		
Item	*Material*	*Quantity*	*Hazardous Substances*

Future State				
Minimize				
Item	*Source of Garbage*	*Qty*	*Recycle/Compost*	*Savings*

Green State					
Elimination					
Item	*100% Biodegradable or Recyclable*		*Elimination*		
	Solution	*Savings*	*Solution*	*Solution Cost*	*Purchase and Disposal Cost*

Figure 14.4 Garbage Waste Elimination Worksheet

Transportation Waste Elimination Worksheet

Activity or Area:	

Current State

Identify	*Measure*	
Item	*Mode*	*Distance*

Future State

Minimize

Item	*Distance*		*Offsetting*
	Solution	*Savings*	

Green State

Eliminate

Item	*Eco-Transportation*	
	Solution	*Savings*

Figure 14.5 Transportation waste elimination worksheet.

Emissions Waste Elimination Worksheet	
Activity or Area:	

Current State		
Identify	*Measure*	
Item	*Type*	*Amount*

Future State				
Minimize				
Item	*Discharge*		*Savings*	*Offsetting*
	Source	*Scrubbing*		

Green State		
Elimination		

Figure 14.6 Emissions waste elimination worksheet.

Biodiversity Waste Elimination Worksheet		
Activity or Area:		
Current State		
Identify	*Measure*	
Item	*Type*	*Amount*
Future State		
Minimize		
Item	*Solution*	*Cost/Savings*
Green State		
Eliminate		
Item	*Regeneration*	
	Solution	*Savings*

Figure 14.7 Biodiversity waste elimination worksheet.

Index

C

H

Z

ext Transpotation Scope ?

mat/prod

inb

IPT

Cust.

mail

courier
inter office mail
usps inbound mail (invoices)
usps outbound mail

people

commuter – biking / driving / carpooling / public
bus travel.
sales fleet cars.

internal transp Scope

inside the bldgs.
offices
plants/DC's

About the Author

Brett Wills has more than ten years of experience in the environmental arena. Brett also has firsthand experience leading teams in the implementation of lean initiatives such as Kanban systems, finished goods supermarkets, and 5S, and has extensive experience in the development and management of ISO 9001 and 14001 systems. He has also been instrumental in facilitating the development and implementation of many health, safety, and environmental initiatives within a manufacturing environment.

Brett began his career as the environmental programs coordinator for an energy-efficient transformer manufacturer, in which he worked closely with various environmental organizations such as Energy Star and Eco Logo to help develop the energy-efficient standard for the dry-type transformer category as well as other environmental initiatives. In a later role as the purchasing and logistics manager, his hands-on leadership approach resulted in a standardized purchasing and materials management process that led to significant reductions in inventory levels and material costs. Brett also developed and implemented the company's highly successful health and safety program. When he was later promoted to operations manager, Brett helped lead the development, implementation, and successful certification of the ISO 9001 and 14001 systems. As plant manger, Brett took on the overall management and continuous improvement of these systems and has led these systems through many successful third-party audits. Also in his role as plant manager, Brett used his extensive environmental knowledge and experience, coupled with his leadership and management ability, to lead the development and implementation of a number of green initiatives, all of which resulted in significant positive environmental and economic impacts.

Brett's blend of hands-on management and leadership experience, along with his knowledge of lean, green, and continuous-improvement concepts, provide him with a unique skill set. His ability to develop a shared vision, and then empower teams and individuals through training and coaching to realize the end goal, produces consistent results.

Brett lives in Toronto and holds a B.A. in economics from Wilfrid Laurier University. In his spare time, Brett enjoys many outdoor activities and has been a nationally certified and practicing ski instructor for the past eight years.